サステナブル社会のまちづくり
海外の実務者との対話から見えて来るもの

編著者：澤田誠二＋著者：大月敏雄＋講演者：国際シンポジュウム講師

Machidukuri in a Sustainable Society
What is effective machidukuri in a globalizing society

Editor & Author: SAWADA Seiji
OTSUKI Toshio + Symposium Panelists

本書は2014年12月1日〜3日の間、東京、横浜、滋賀で開催された「サステナブル社会のまちづくり」国際シンポジュウムの記録と関連資料をもとに記述、編集した。

国際シンポジュウムの概要

主催	明治大学サステナブル建築研究所＋明治大学国際連携本部
	横浜交流会実行委員会
	滋賀県立大学・地域共生センター
後援	(一社)団地再生支援協会
	UR都市機構
	国土交通省
	NPO都市計画家協会
	(一社)東京自治研究センター
	関西大学・地域再生センター
	NPO日独文化コミュニティ
協賛	明治大学
	滋賀県立大学
	(公財)大林財団
	(株)合人社計画研究所
	NPO団地再生研究会
	YKK-AP株式会社
	積水ハウス株式会社
	大和ハウス株式会社
実施	「サステナブル社会のまちづくり国際シンポジュウム」実行委員会

■ How the Machizukuri symposium came up

This symposium is an inheritance of the international seminar at Meiji Univ. in 2011 and the JGCC symposium in 2013. These German-Japanese comparative studies were programs to explore the problems which seem to be beyond that of refurbishment of large housing complexes.

The contents of this book are mainly summaries of the reports at the International symposium on Community Development in Globalized Sustainable Society, Tokyo 01., Yokohama 02., Shiga 03. Dec.,2014.

Outline of the symposium

Organized by Sustainable Building Research Unit
and International Corporation Div. at Meiji University,
Yokohama Exchange Committee,
Regional Symbiosis Center at University of Shiga Prefecture.
Supported by Housing Refurbishment Association,
Urban Renaissance Corporation,
Ministry of Land Infrastructure and Transport (MLIT),
Japan Society of Urban Planners (JSURP),
Autonomy Tokyo,
Kansai University,
Japan-Germany Culture Community
Sponsored by
Meiji University,
University of Shiga Prefecture,
Obayashi Foundation,
Gojinsha-Keikaku-Kenkyujo,
Housing Refurbishment Study Group,
YKK-AP Corporation,
Sekisui House Corporation and
Daiwa House Corporation.

東京会議、横浜会議、滋賀会議の概要

東京会議	12/1(月)	午前	明治大学アカデミーコモン 100名
		午後	(同上)
横浜会議	12/2(火)	午前	汐見台団地の紹介と現地視察 80名
		午後	神奈川県産業振興センター
滋賀会議	12/3(水)	午後	滋賀県立大学・交流センター 90名

Conferences

Tokyo	12.1. Mon	morning	Meiji University 100
		afternoon	Meiji University 100
Yokohama	12.2. Tue	morning	Visit to ShiomidaiHousing
		afternoon	KIP Conference C. 80
Shiga	12.3. Wed	afternoon	Univ. of Shiga Pre. 90

はじめに

サステナブルが何を指すのかを問われると、今の時代、すぐに答えられることはけして多くない。一方で「まちづくり」に対しては、様々な立場の人から、それぞれの専門に関わる答えが返って来る。

自分が「サステナブルなまちづくり」というテーマを意識したのは、15年前にドイツ・デッサウの団地再生国際会議に参加した時だった。住宅団地がテーマで、1960年代からその供給システムの開発に関与していたからだ。住宅の多量供給が始まって以来 40年経ったデッサウ会議当時のヨーロッパの団地は老朽化が著しかった。特にドイツは東西統合（1989年）のため住民が流出し、団地には多量の空き家が発生していた。そうした様子を見て、まちづくりという行為が思いのほか短命だと感じたのだ。会議では老朽化した団地をどのように再生するのが良いかについて議論され、建物の老朽化だけでなく団地開発以来の社会変化を含めた包括的なとらえ方が不可欠だという結論になった。

この時から、まちづくりの行為の進化の経緯と、まちづくりをとり巻く地域や社会の変化とを合わせて捉えるために年表をつくるようになった。この年表の上に、団地再生のプロジェクトを位置づけて、プロジェクトの目標と目標達成のためのアプローチを分析すると、「サステナブルなまちづくり」の意味が明確になると考えた。

今の時代、サステナビリティつまり持続可能性といえば、エネルギー、自然環境から会社経営のあり方まで広い分野のことを差す。しかし本書では"サステナブルなまちづくり"を、しっかりした土地の上に人々の生活や産業あるいは文化を支える住環境をつくること、その住環境は生活や産業・文化の変化・発展に対応することが出来て、出来るだけ長く使えるようにすること、と設定している。

Introduction

Not many people today can easily give an answer to what sustainability means. However, when I ask people about "Machidukuri (planning, design construction and management of built-environment)," I find that many people coming from all walks of life are able to give me an answer tied to their professions.

I first began giving "sustainable Machidukuri thought 15 years ago when I attended an international conference in Dessau, Germany on the refurbishment of housing complex. I attended the conference, which focused on residential housing estates, because I had been working on the development of the housing supply system since the 1960s. At the time of the Dessau Conference, 40 years had passed since a large number of housings had become available in Europe, and the deterioration of these housing estates had become a serious issue. The situation was particularly dire in East Germany, where many residents had moved out following the unification of East and West Germany (1989) and many houses were left uninhabited. Examining this situation, I realized how short-lived the results of the Machidukuri activities were. Strategies for the deteriorating housing estates refurbishment were discussed at the conference, and we reached the conclusion that it was vital to examine comprehensively not only the aging of the buildings themselves but also the societal changes after the development of the housing estates.

After this, I began creating a timeline to examine both the evolution of Machidukuri and changes in the region and society that pertain to Machidukuri activity together. I believed that if I were to contextualize the housing estate refurbishment projects on this timeline and analyze the project's goals and the strategies necessary to achieve that goal, it would illuminate what "sustainable Machidukuri actually means.

Today, the word sustainability is used in a wide variety of contexts such as energy, natural environment, and business management. However, in this book, I have defined the term "sustainable machizukuri to describe the act of creating a living environment in a stable region to support people's lifestyles, industry, and culture so that it is flexible enough to be adapted for societal changes and can be used for as long as possible.

1800 ⋯⋯ 1900 ⋯⋯

近代まちづくりの発端　Beginning of Machidukuri

日本の農家（茨城県平野家）
Japanese farmhouse
写真提供: 日本の民家、山渓カラーガイド

ドイツの農家（中庭）
German farmhouse

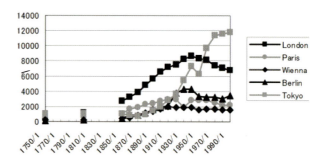
ヨーロッパと日本の都市形成－19世紀〜20世紀（大都市人口）
Population change of big cities in Europe and Japan, 19th 〜 20th century

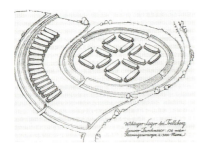
トゥレボルク（ヴァイキングの居留地）
Φ136m、1,300人居住、11世紀
Wikinger Lager bei Trelleborg, Denmark
提供: 永松栄

住宅はまちづくりのもと
近代まちづくりは産業革命から
－産業の発展が近代都市と国家を発展させた
－都市、産業地帯、農地を計画する制度、技術が発達した
新建築・まちづくりの材料、技術の開発・普及
－啓蒙システム－万国博覧会－新材料、新技術、新市場
研究・教育システム－バウハウス

- Machidukuri begins with building home & house
- Machidukuri of Modernization era
  Supported by industrialization, for population concentration to cities
  Machidukuri system is to control urban, industrial, and farming area
  Measures to develop and popularize new technologies and materials・
  Popularization: exposition system・Research, development and education / training

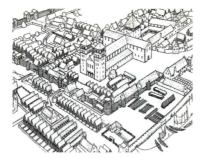
シュトラスブール（フランス）の中心部 11世紀
Strasbourg, France, City center, 11th century
提供: 永松栄

ケルンのアグネス教会中心の新市街地開発　1900年ころ
Development of new district with Agnes Church,
Cologne Germany, around 1900
出典: "Leben/Kultur/Stadt　1840-1900",
Greven Verlag, Köln

19世紀　ルール工業地帯の風景　英文
Germany Ruhr Industry Area, 19th century
出典: North-rhein-Westfalia State

19世紀　ルール工業地帯の風景　英文
Germany Ruhr Industry Area, 19th century
出典: Northrhein-Westfalia State

● バウハウス発足 1929
BAUHAUS Establishment

明治政府　Meiji Government Establishment 1968
廃藩置県　The abolishment of feudal domains
　　　　　followed by the establishment of prefectures 1971
産業地帯と大都市の形成
Development of cities and industries

1950
戦災からの復興、社会成長に向けた助走
インターバウ国際建築展（1957年、ベルリン）

Run-up for the Social Development
Interbau Housing Exhibition, Berlin, 1957

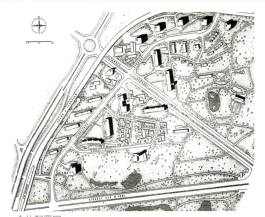

全体配置図
General plan

写真提供：大坪明

ティアガルテン西部に著名建築家デザインの集合住宅が並ぶ　　写真提供：大坪明
Panoramic view, houses designed by famous architects

グロピウス・ハウス
Gropius-House

アールトー・ハウス　　写真提供：大坪明
Aalto-House

ニーマイヤー・ハウス　　写真提供：大坪明
Niemeyer-House

● 西欧の団地開発
　Housing developments in the western Europe
● フランス、オランダ、デンマーク、スウェーデン
　France, Holland, Denmark, Sweden

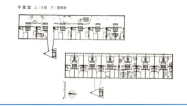

住宅供給体制づくり（公団、公社、公営住宅＋金融公庫）
Japan Housing Corporation, Housing Supply Corporations,
Housing Finance Corporation

1960
社会,経済の成長豊かさを求めて Social and Economic Development
建築家からの提案の時代 Architects' Ideas "City of the Future"
Federal Environmental Protection Law ドイツ: 連邦自然保護法（1967年）

フラー：ドームでニューヨークを覆う
Fuller dome

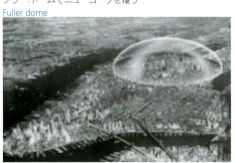
フラードーム 二面立体トラス構造
he dome system, Fuller

クック：ウェストコーストの楽しいまち
West Coast city, Cook
出典：古山涼（修士研究）

黒川：ヘリオックス（都市空間システム）
Heliox system, Kurokawa

丹下：東京計画（海上都市）
Tokyo 1964 plan, Tange

東京、東京駅周辺
Tokyo, Central Station area
写真提供：TEKES, Finland

京浜工業地帯
Keihin Industry Area

- ハブラーケンサポート理論
  "house(Support) / community / man", Habraken

- ティッシュ・サポート・インフィル
  Three Levels of built-environment: Tissue / Support / Infill

- 安保闘争 1960年　The Japan-U. S. Security Treaty
- 都市計画法　Urban Planning Law
- 東京オリンピック 1964年　Tokyo Olympic Games 1964
- 規格化、部品化、標準レイアウト Industrialization of housing: layout, house type, building component
- 大規模住宅団地の建設　Large housing estate development

1970
社会成長、都市化の進行
Social development, Urban Development

ミュンヘン・オリンピック（1972）
Munich Olympic Games 1972

パリ、中心地区
Paris, France, Central district1987 vivre paris, mengre

ベルリン、中心部の船着き場
Berlin, Harbor at the Spree River

カルマール（スウェーデン）の団地
Karmar, Sweden, Housing at

東京、西から新宿副都心をのぞむ
Tokyo, A view of Shinjuku from the west

ルール工業地帯　夜景
Ruhr indusrtry area in the night

ルール工業地帯　再開発中
Ruhr indusrtry rea in re-development
写真提供: Northrhein Westfalia State

●東欧の団地開発
Many "Panel" housing Development in eastern and central european countrives;
東独、ポーランド、チェコスロバキア、ハンガリー、ソ連
GDR, Poland, Checoslovakia, Hungary, Soviet Union

千里ニュータウン
Senri New Town
写真提供 UR都市機構

千里ニュータウン・まちの風景
Senri New Town – Town Scapes
写真提供: 大坪明

- ●大阪万国博覧会　World Expo Osaka 1970　●プラザ合意　PLAZA Agreement
- ●都市計画法・都市再開発法　Urban Planning Law / Act Revision　Urban Development Law / Act
- ●多摩ニュータウン入居開始　●住宅産業・住生活サービス産業
   Tama New Town Development　　Development of Housing Industry: Hard & Softsystem

1980　　　　IBA エムシャーパーク　IBA Emsher Park

経済のグローバル化　　情報社会化
Globalization of economy　Information-oriented society

プロジェクト・エリアの地図 / 対象エリアは6区分
Map of the redevelopment area, subdevided into 6 areas

ガーデンシティ的な炭鉱労働者ハウジングの正面、全面保全
Front view of a large coal miner housing complex, mostly preserved

ガーデンシティ由来の炭鉱労働者住宅の再生―各戸に菜園付き
Garden City style coal miner housing, with vegetable garden

産業排水が溢れがちな場所を再自然化し、
遊水池機能を持たせている。（地中に汚水管を埋設）
Some areas where rain water flood pond was designed
(Industrial drainage piping underground)
写真提供: North-rhein Westfalia State

練炭工場跡地に開発されたエコ・ハウジング、中庭は遊水池
Eco-housing redeveloped on the site of briquette factory

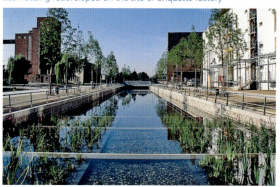

運河沿い水辺環境を整備し倉庫等リノベーション住宅を建設
Canal-front landscaping and store house-renovation housing

BAエムシャーパーク地域再生プロジェクト
・NRW州北部を東から西に流れるエムシャー川と緑地帯とを修復し、
　"水と緑のネットワーク"による公園（800km²）をつくる。
・公園に州、地元、企業による100プロジェクトを遂行。
・成長なき時代にハイクオリティの生活・産業・環境を実現。
―タウンセンター再開発
―ハウジング
―産業施設の再生
―インフラの再生
―ランドスケープの再生

IBA Emsher Park regional development, 1988-2000
Brownfield redevelopment
Park consists of "green zone" and "water line" (Emscher River)
100 projects, of public, private developers, for project concepts:
・Town center
・Housing
・Industry
・Infrastructure
・Landscape

● 地区計画制度　　　　　● 地方分権化（四全総）　　　● バブル景気

Building Regulation Plan　　**4th Development Plan**　　**Bubble Economy**

急速に進む産業構造の変化
Structural change of industries / Brownfield redevelopment

ベルリンの壁崩壊・ドイツ東西統合
The end of the Cold War / German reunification

製鉄所にできた屋外ステージ / Outdoor Theater in ironworks

ボタ山に設けられたピラミッドの展望台 / Pyramid on top of slagheap

林を抜ける歩行者ブリッジ
Pedestrianbridge through trees

内陸港にビジネスセンター、ハウジングを開発
A large canal harbor re-development: business centers and housings.

19世紀の産業施設に残る
ユーゲントシュティールの玄関
Jugend style design entrance of assembly hall from 20 century.

ガラス壁の R&Dセンター /
Glass wall of a research center.
写真提供: Northrhein Westfalia State

地域全体がランドスケープ・パークとしてデザインされた
Large area was designed as "landscape park".

● ハウスメーカーなど住宅産業が躍進
Rapid progress of house maker (prefab house industry); ca 10,000 units / year, 1985

ハウジング産業の構造、1985
Composition of housing supply industry, 1985
全体 (10,000戸) の半分 (上部) がデベロッパーによる主にRC造集合住宅、下部が 戸建住宅。右部分がプレファブ住宅と部品

1990

ライネフェルデの団地再生　Refurbishment Leunefelde

社会変革　ドイツ都市再生政策（Stadtumbau Ost Policy）　京都議定書
Social Change　Renovation of cities and housing estates in the East　Kyoto Agreement

ライネフェルデの位置
Leinefelde at the center of Germany

団地再生のマスタープラン
Project master plan

中央街区にできた広大な中庭と日本庭園
Large courtyard of the central quarter, with Japanese garden

昔は小さな町だった
Small town, around 1930

再生前の住宅団地
View of the housing area before the refurbishment, 1989

解体PCパネルを利用したデザイン
"Sun Shine" kindergarten with reuse of PC panels

全景：手前が団地再生エリア、後方に工場、周囲は田園地帯
Panoramic view: industry area behind the housing area, 2005
写真: Stadt Leinefelde

中心部の日本庭園からの眺め、
View from the Japanese garden in the courtyard

- 都市計画法改正（まちづくり三法）
  Improvement of Town Planning Act
- 中心市街地活性化法（まちづくり三法）
  Chushin-Shigaichi (Downtown) Revitalization Act

2000

EUサステナブル社会づくり本格化
Sustainable machidukuri promotion in full-scale

ハノーバー「環境」万博
"Environment" EXPO Hannover 2000

整備後の中央街区の中庭／
Central quarter courtyard, 2005

長大な棟の"間引き"によってアーバンビラに
Urban Villas, which were made from a long apartment building
提供: NEUBAULAND

減築しテラスハウス化した棟と
コーナーを埋めるカフェ
One floor-reduced and balcony-added house next to new café
写真: Stadt Leinefelde

団地再生プロジェクトの最終段階: ランドスケープデザイン
Final stage of the refurbishment –andscaping of the area
提供: Stadt Leinefelde

● EUスロイロ団地再生システム開発プロジェクト
  Sustainable Refurbishment Europe Project,
  2000-2004
● CIB W104 Open Building Implementation, 2,000

5階建てを4階に、解体中の
アーバンビラ
Fifth floor was removed and middle sections were pulled extracted.

NEXT21 – "スケルトン・インフィル"集合住宅システム
"Lonifestructureandchangeable component" housing
出典:, Domus 2003

● 大規模小売店舗立地法（まちづくり三法）
  Act on the Measures by Large-Scale Retail Stores for Preservation of Living Environment

● NEXT21 長期実験プロジェクト、大阪ガス
  "Base-building & Fit out" i.e. Support / Infill system experiment, by Osaka Gas
● ストック再生・利活用 Stock-Renovation Promotion

2010
エネルギー改革の本格化 Energy Reform
少子化、高齢化、地域格差 Population Decline, Aging, Regional Difference
● ドイツ原発削減を決定　Decision on Nuclear Power Plant

歴史的転換期に入ったドイツの都市開発
○人口の集中と都市の拡大　○都市拡大どゝばらげが進む　○都市に"抜け地"ができて縮退

密集エリアから減築エリアまで：都市改造の対象エリアは5種
○安定化エリア　○潜在エリア　○要保全エリア　○虫食いエリア　○要減築エリア

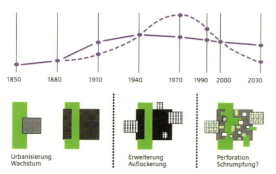

"都市改造政策"の考え方ー人口動態と都市計画コンセプト
Urban development scheme introduced in 1990s "historical change", "Stadtumbau".
1850-1925 Population concentration & urban growth
1925-1980 Urban growth & reducing tension
1980-2030 Perforation & declining
出典: ARGE STADTUMBAUOST

"都市改造政策"まちづくりの様々な手法
Various "Stadtumbau" strategies: Combination of consolidation-, potential-, reserve-, perforation- and reduction-areas

東京、2010、再開発の進む丸の内
Tokyo, Tokyo station and redevelopments in Marunouchi
写真: UR都市機構

ベルリン、2005、ポツダム広場から東ベルリン中心部を臨む
A view of east Berlin from Potzdam plaza

金山町（山形県、中山間地）まちづくり、
建築学会業績賞、2002
Kaneyama Town, Yamagata Pref.
JIA Award　写真: 金山町、住吉洋二

日本最大の団地開発・
多摩ニュータウンの現状
The largest Japanese housing estate Tama New Town, 2010
写真: UR都市機構

街並み
Streetscape

ひろば Market
Communication area

● 東日本大震災（2011.03.11.）
The Great East Japan Earthquake

● 世界建築家連盟東京大会（2011.10.）
UIA Tokyo Congress

● 団地再生モデルプロジェクト（2013~）
Introduction of the Housing Refurbishment Model Development, MLITT

国際シンポジュウムで取上げたサステナブル社会のまちづくりプロジェクトと促進策
Machidukuri Projects & Measures in Japan and Germany introduced in the Symposium Study

近代史におけるまちづくりの発展と、グローバル化への対応を考慮してドイツと日本のサステナブルなまちづくりプロジェクトとその促進策とが選ばれた。

日本の事例の多くが老朽化住宅団地の再生プロジェクトなので、それらが広域レベルでどのように位置づけられているかを合わせて紹介できるように考えた。

ドイツでは、ドイツ統合以来の社会システムの変革がまちづくりにも影響しているので、連邦制社会の中で、その変革がどのように進んでいるか浮き彫りにできるプロジェクトを選んだ。

■日本のプロジェクト事例（住戸数・住人数 / エリア面積）

□団地再生プロジェクト
01:洋光台		4,373戸 / 209ha
02:汐見台団地		3,516戸 / 75ha
03:貝取豊ヶ丘団地		1,302戸 / 19ha
04:観月橋団地		540戸 / 3ha
05:鎌倉グリーンハイツ		310戸 / 5ha
06:たまプラーザ		1,214戸 / 13ha
07:相武台団地		2,500戸 / 31ha
08:浜甲子園団地		4,600戸 / 31ha
09:多摩平の杜		144戸 / 18ha
10:男山団地		4,600戸 / 185ha

□都市再生・地域再生プロジェクト
21:横浜市		3,690,000人 / 427km²
22:滋賀県		1,390,000人 / 4,017km²
23:富山市		425,000人 / 1,242km²
24:黒部市		42,000人 / 426km²

□20世紀後半に始まる伝統建築環境再生まちづくりの事例
25:金山町（山形県）（注） 　　　　　6,175人 / 162km²

■ドイツのプロジェクト事例、そのまちづくり促進策

□サステナブル社会づくりのプロジェクト事例
31:イノベーションシティ
　　ボトロップ市、NRW州　　　　　120,000人 / 100km²
32:IBAチューリンゲン州
　　地域再生事業・東独　　　　2,334,000人 / 16,172km²
33:ゲーラ2030まちづくり
　　プロジェクト・東独　　　　　　100,000人 / 152km²

□1990年代に始まった団地再生・まちづくり事例
41:ライネフェルデ市（注）
　　チューリンゲン州　　　　　　18,5000人 / 97km²

参考:
25:金山町について：山形県最上郡。2002年、建築学会業績賞受賞
11:ライネフェルデ市について：サステナブル社会のまちづくりードイツ・EUの実務に学ぶ、2912、明大出版会叢書

Aking historical developments of Machidukuri in modernization eras, and its possible contribution to the globalized societies into consideration, following projects were selected.

As for the Japanese projects, they are to be introduced in relation to the programs in higher levels.

A radical change of German traditional federal system since the reunification made substantial change of Machidukuri system. The selected projects should show that clearly.

■ Projects in Japan (dwelling units or population / land area)

□ Housing Estates Refurbishment Projects;
01 Yokodai		4,373 DU / 209ha
02 Shiomidai		3,516 DU / 75ha
03 Kaitori-Toyogaoka		1,302 DU / 19ha
04 Kangetsukyo		540 DU / 3ha
05 Kamakura Green H.		310 DU / 5ha
06 Tama-Plaza		1,214 DU / 13ha
07 Sobudai		2,500 DU / 31ha
08 Hamakosienn		4,600 DU / 31ha
09 Tamadaira-no-Mori		144 DU / 18ha
10 Otokoyama		4,600 DU / 185ha

□ Urban, Suburban, Regional Development Projects
21 Yokohama City		3,690,000 p./ 427km²
22 Shiga Prefecture		1,390,000 p./ 4,017km²
23 Toyama City		425,000 p./ 1,242km²
24 Kurobe Cit		42,000 p./ 426km²

□ Machidukuri project started in 1980ers in local tradition
25 Kaneyama (Yamagata)　　　　6,175 p./ 162km²

■ Projects, Promotion Measures in Germany

□ Community Development for Sustainable Society
31 Innovation City Bottrop　　　　120,000 p./ 100km²

32 NRW / West Germany 132 IBA Thüringen State
　　East Germany　　　　2,334,000 p./ 16,172km²
33 Gera 2030 City Project
　　East Germany　　　　　100,000 p./ 152km²

□ Housing Estate Refurbishment Projects since 1990s
41 Leinefelde　　　　　　　18,500 p./ 97km²
　　Thüringen State

01~10：団地再生プロジェクトについて：
（一社）団地再生支援協会に問合せのこと

序文　サステナブル社会ー脅威からの解放の先にあるもの
Preface　Sustainable Society – getting liberated from the threats and beyond

大月敏雄
OTUKI, Toshio

■サステナブル社会を「脅威」から考える

　本書に登場するドイツからの論客、H.シュトレーブ氏のいうように、「こうなっていなければサステナブルではない」という定義はおそらくないだろう。だからこそ、サステナブル社会とは一体なんなのかということは、日々問い続けなければならないのだろう。ここではまず、人間の生活に対する脅威について考えてみることを通して、サステナブル社会とは何ぞやということを、その根源に立ち戻って考察してみたい。

　人間の生活に対する主要な脅威について、歴史的段階的に人類が直面してきたであろう順番を考えてみると、大まかに「飢え」「病気」「自然災害」「戦争」「科学」「人口減」のように描けるのではないだろうか。

■飢えの脅威

　「飢えの脅威」はすなわち貧困の脅威であり、食糧の確保に始まって、ひいては経世済民という意味での経済がきちんとしていることが、飢えの対策の基本となろう。これを建築やまちづくりの世界に引き寄せると、社会状況が激変した場合（例えばエネルギー革命で石炭が要らなくなったとか）でも、ある程度地域的な自給力を蓄えておける機能や空間の再構成が必要ということになる。また、「食」と「職」の問題は深く結びついており、産業をどのように地域の中で構成し続けていくべきかという、古いようであっても極めて現代的な課題とも結びついている。また逆に、「飢えの脅威」から逃れようとするあまりに、「儲けすぎる」ことを善とする風潮が、あたかも人間の自然な行動だとして理論化され、それが過剰な資本の偏りや過剰な世界的資源争奪を生むことによって、「戦争の脅威」を引き起こしたりするのである。

■病気の脅威

　次に「病気の脅威」も、人類が生命である限り闘い続けなければならない課題である。建築やまちづくりにおいて特に課題となったのは、19世紀に先進国で深刻になったスラムの発生とその予防についてである。近代都市計画の根幹はこれによって成り立っている。コレラやペストといった伝染病は、人間の貴賤を問わずに平等に襲いかかるゆえ、貴族が己の身を守るためにハウジングや衛生的な都市整備に税金を費やし始めたのだと、私は理解している。この問題は、今でも「スラムの問題」として地球の大部分の都市問題の核をなしている。また、サーズやマーズ、エボラ出血熱といった病気群も相変わらず我々の脅威である。さらに、近年東京の都市公園でデング熱を媒介する蚊が見つかったために、公園が出入り禁止となった。気候変動はかような形で、一見、衛生的によく制御されているかに見える近代都市において、「病気の脅威」をばらまく可能性があることを認識しなければなるまい。

■自然災害の脅威

　そして日本の建築やまちづくりを語る上で外せないのが「自然災害の脅威」である。日本の地理的条件によって生じる地震、津波、噴火といった自然災害は、根本的に防ぎようがない。でも、似たような自然災害を幾度も被ることを通して、日本の建築づくり、まちづくりは少しずつではあるが、進展してきたことも間違いない。東日本大震災以後に主流

■ Thinking about Sustainable Society from a "threat" perspective

As the German commentator, who appears in this book, Mr. H. Straeb says, there is probably no rule that states "if it isn't done this way, it isn't sustainable". For this reason, we should continue to question what a sustainable society really is. First, by thinking about threats to human life, and going back to basics, let us examine what a sustainable society is.

If we try to identify the major threats to human life and line them up more or less in the historical order in which they may have appeared, one can roughly break them down into "famine", "disease", "natural disaster", "war", "science", and "population decline".

■ Threat of famine

The "threat of famine" is in other words the threat of poverty, and the fundamental form of countermeasure to this threat starts with the securing of food, and eventually leads to a government and economy that can provide for the needs of a given people. If we carry this over to the world of architecture and Machidukuri (creation of a community), there is a requirement to reconstruct spaces and functions so that a community can at least to some extent be self-sufficient in the event of a major change to our social situation (for example, there being no more need for coal due to the energy revolution). Additionally, there is a profound connection between the issues of "shoku (meaning food)" and "shoku (meaning work)" which is also linked with the age old yet exceedingly contemporary problem of how we should continue to develop industry in the community. Conversely, we now live in a world where greed is considered a part of human nature and even a positive attribute in our efforts to escape the "threat of famine", which can then lead to the "threat of war" by creating an excessive polarization of capital and a struggle for access to resources.

■ The threat of disease

The "threat of disease" is another problem that homo-sapiens will have to continue to face as long as we exist as living creatures. The major issue in terms of architecture and Machidukuri was the appearance of slums, which became a serious problem in developed nations in the 19th century, and policies to eradicate them. This has been the basis of modern town planning. My own view is that it was to protect themselves that the aristocracy began to spend taxes on housing and reorganization of urban infrastructure to improve sanitary conditions, as infectious diseases such as cholera and plague spread and killed thousands irrespective of class or wealth. This "slum issue" is still at the core of urban problems across most of the world today. Additionally, diseases such as SARS, MARS, and Ebola continue to be a major threat to humanity. In recent years, the discovery of mosquitos that were acting as carriers for dengue fever in urban parks in Tokyo led to the temporary closure of these parks. It is important to recognize that climate change can potentially spread the "threat of disease" in such ways even in modern cities that appear at first glance to be hygienically well controlled.

■ The threat of Natural Disaster

In discussing architecture and Machidukuri in Japan, one point that

となった、「防災から減災へ」という流れは、建築も土木構築物も、決して強すぎてはならず、弱すぎてもならず、多重に、そして多様に考えられた「生き延びるためのシナリオ群」の中で人々が生き続けることをどう保証するかという点を重視し始めた。しかし、自然災害がきっかけとなって起きる原発事故は明らかに人災であり、深い反省さえあれば繰り返すことが避けられるはずであり、ドイツではこれに成功しつつあるが、日本ではなかなか止めることが出来ないでいる。さらに、これまで人類が記録をしてきた範囲内では起こらなかったような気象の変化が大きな脅威をもたらしつつある。近年見たこともないような降雨量や風速、暑さ。こうした気象変動もコントロールできるのではないかという期待が、CO_2削減を世界規模で流行させており、それがひいては建築やまちをつくっていくことに大いなる制約を課しつつある。気象変動はこの点において、地震や津波とは異なるインパクトをもっているのである。

■戦争の脅威

次に「戦争の脅威」。戦争は個人にとってみれば、己の意思に反して人を殺し、また己の意思に反して人から殺されることである。都市にとってみれば、その都市の存在の意図に反して、破壊されることである。この脅威が生じる主たる原因は、過剰なまでの「飢えの脅威」からの逃亡であろう。また、ある戦争が原因となって他の戦争を引き起こし、それがもたらす復讐心が民俗の意思決定の一要因ともなり、文化や宗教という形で人間の行動様式を固定化する面もある。建築やまちづくりにとって最大の敵であるこの戦争という脅威は、我々の活動によってダイレクトに軽減できるものではないかもしれないが、我々は「飢えの脅威」をうまくコントロールすることを通して、一生懸命戦争の脅威を撲滅しなければならないのだろう。

■科学の脅威

20世紀中に起きた二つの地球規模の全面戦争はまた、新たな脅威を人類に気づかせてしまった。「科学の脅威」である。第一次大戦においては、産業革命の延長として登場した機械的大量殺戮兵器が、そして第二次世界大戦においては、普通の人の目にはその因果関係が理解できない形で科学が寄与した核爆弾が、「戦争の脅威」を超える「人類絶滅」の脅威を経験させた。これが「科学の脅威」とも呼べるしろものであるらしいことは、大戦後の様々な地球規模の人類生存の脅威として指摘されてきた事象にも見ることができる。1962年のレイチェル・カーソンの『沈黙の春』はそのことを解り易く人類に訴えた。事実、それと相前後するように、日本では水俣病をはじめとする公害問題が相次いで取りざたされた。経済活動の思考範囲を自己都合的に矮小化して設定された領域の中では最適解であったはずの科学の応用がもたらした化学物質が、外部不経済という形で、想定範囲外で暮らす人々に生存の脅威をもたらしたのである。科学の脅威は、実験室的思考が想定している範囲外のことについて無頓着であるという意味で、外部不経済から常にしっぺ返しを食らっている。福島第一原発の事故もその一つである。また、そこまで因果関係が明白でなくとも、CO_2による地域温暖化という仮説を、いまだに人類は捨てきれずにいる。CO_2という外部不経済は排出権の売買問題という形でいま、別の科学によって、その制御の可能性が議論されつつある。科学の脅威を科学の応用によって

cannot be overlooked is the "threat of natural disaster". The natural disasters such as earthquakes, tsunamis and volcanic eruptions that occur in Japan due to its geographical situation are basically unpreventable. It is, however, also true that Japan's architecture and Machidukuri have slowly but surely evolved through suffering similar natural disasters many times. The notion of "reduction rather than prevention of the effects of natural disasters" that became mainstream following the Great East Japan Earthquake has placed more emphasis on how to guarantee people's survival within the diversely developed "Survival Scenarios Portfolio" and the notion that architecture and civil engineered structures should not be too strong or too weak. However, the nuclear disasters that occur due to natural disasters are undoubtedly man-made disasters, and should be preventable with sufficient forethought, and while Germany is showing signs of success, Japan is still struggling. Furthermore, changes in the climate that have previously never been experienced in the history of mankind including rainfall, high winds, and extreme heat, could become a further threat. The anticipation of potentially controlling such climate change has popularized CO_2 reduction on a world scale, and could in turn impose restrictions on the construction and development of towns. In this way, climate change has a different impact from earthquakes and tsunamis.

■ The threat of war

For the individual, war is to kill and be killed against one's will. For a city, it is to be destroyed despite its wish to exist. The main cause of this threat is most likely the escape from the "threat of famine" at an extreme level. Additionally, one war becomes the catalyst for further wars, and the desire for revenge caused by this becomes a factor in the decision making process for a nation or race of peoples, thus leading to rigid behavior patterns in terms of culture and religion. The threat of war, the greatest threat to architecture and machizukuri, cannot be reduced directly by our actions alone but, by controlling the "threat of famine", we must do our best to eradicate this threat.

■ The threat of science

The two major World Wars that took place in the twentieth century awakened humankind to a new threat, the "threat of science". In the First World War, machines capable of mass destruction created as an extension of the industrial revolution, and in the Second World War, nuclear bombs, given to us by science (though the causality of this is often not understood by ordinary people), led us to experience the "threat of extinction" above and beyond just the "threat of war". That this is the so-called "threat of science" is clear from the many post-war worldwide threats to the existence of humankind. The 1962 book "Silent Spring" by Rachael Carson clearly set this out for the entire human race. As a specific example, around the same time in Japan, many pollution problems such as Minamata disease were causing a stir. Chemicals created by the practical application of science, regarded as the perfect answer to all our problems in a microcosmic world where personal gain and the needs of economic activity trivialized everything else, became a life threat to people living outside this assumed sphere in the form of external diseconomies. The threat of science is always being knocked back by external diseconomies due to its lack of interest in anything other than what can be predicted within the confines of a laboratory

乗り越えようとしているのである。また一方で、科学は「有限な資源」という限界にも挑戦しつつある。この問題をセンセーショナルに問うたのは、1972年に発表されたローマクラブによる『成長の限界』であり、その翌年に起きた第一次オイルショックによって、事の信憑性が裏付けられた。そして、それを乗り越えるためにクリーンエネルギーといった科学が、さらに応用されつつあるのだが、自然科学にしろ社会科学にしろ、科学はある想定エリア内で理論化されるために、想定していないエリアからのしっぺ返しという脅威を決して忘れてはならない。そうした科学の限界を乗り越えようとするイメージの共有の一例として、ジェームズ・ラブロックによるガイア仮説を挙げておこう。地球上に想定外のエリアを想定しないというこの魅力的なアイデアは、科学の脅威から逃れる安心感をもち得るための「一枚の絵」である事には間違いなかろう。

■人口減の脅威

そして、地球上で一番深刻に日本が直面しつつある課題が、「人口減の脅威」である。確かに、人口が減りすぎると本当に絶滅してしまうわけで、その事は人類のせいで破滅に導かれたたくさんの生物種の最期の様子を見ても容易に想像がつく。というわけで、人口増は確かに切実な課題であるが、とりもなおさずコンパクトシティ化を進め、とにかくインフラの維持にお金がかからないようにというのが、今の日本政策の向かっているところである。しかしながら、比較的過密に都市を構築してきた日本人にとって、もっと低密に快適に暮らす当選択肢を考える余裕も、同時にあってよい。コンパクト化の議論は、あらゆるインフラは100％行政の仕事であるという前提のもとに生まれている。例えば、長野県下條村では村が資材を支給し、住民が道普請をすることでインフラコストの低減を図っている。もちろん限界があろうが、100％税金に頼らないインフラの維持はどこまで可能かといったことに、より意識的にチャレンジしていってもいいだろう。

■部分最適解から全体最適解へ

以上のように、サステイナブル社会を考えて行く時の、いくつかの根源的問題をひもといてみたのだが、本書で紹介されているドイツの最新のまちづくりの様子を伺ってみると、ここで指摘した脅威について、実に包括的にアプローチされているように感じる。ドイツの事例は、地方自治体以上の広域レベルのものが多く、日本の事例よりも大きい。単に面積が大きいばかりではない。アプローチ自体も総合的かつ統合的なのである。この点でいえば、日本で論じられているのは団地最適解、あるいはプロジェクト最適解といってもよい、部分最適解なのである。もちろん、地方自治体単位で取り組まれている富山市のような例もないではないが、これが圧倒的に少ない。これは、日本におけるプロジェクトというものの認識にかかっているのだと思う。

ドイツの事例のほとんどは、ある社会的課題に対して国家、あるいはそれに準ずる地域レベルで、事業コンペを通して厳選されたプランが先導的な役割を担うように仕組まれている。こうして、選ばれたプロジェクトは広域レベルの問題解決の代表選手として、その失敗も成功も、他のプロジェクトのお手本になるように仕組まれている。だから、必然的に総合的、統合的になっている。単なるアイデア実践ではない。言葉を換えれば、部分最適解を統合して広域の最適解を導くためのシス

experiment. The disaster at Fukushima No.1 Nuclear Power Plant is a case in point. Furthermore, even though the causal relationship isn't particularly clear, humanity still cannot set aside the hypothesis that CO_2 causes global warming. The potential control of the external diseconomy that is CO_2 is currently being debated through the use of another kind of science in the form of the sale of emission credits. We are trying to overcome the threats posed by one aspect of science through the application of other scientific concepts. On the other hand, science is also trying to challenge the limitations imposed by "finite resources". The issue of finite resources was sensationally brought to the fore by the 1972 book "The Limits to Growth" commissioned by the Club of Rome, the credibility of which was supported by the first oil shock that occurred the following year. To overcome this problem, scientific concepts such as the notion of clean energy are being pressed into service, but, as science, whether natural or social, starts from a basis of theoretical assumptions, we must be conscious of the danger of having it all thrown back in our faces because of a theoretical factor we failed to take account of. As an example of science trying to overcome its limits, let me mention James Lovelock's Gaia Hypothesis. This attractive idea, which states that everything on earth has to be taken into consideration, is without a doubt "one scenario" for potentially escaping the "threat of science".

■ Threat of decrease in population

Finally, in global terms, potentially the most serious issue Japan faces is the "threat of decrease in population". If the population decreases too much, the obvious outcome is that we as a species will become extinct and one can readily imagine this situation by observing the many species we have driven to extinction. While population increase is a pressing issue, the current direction that Japanese policy is taking is to continue creating more compact cities so as to reduce the cost of maintaining infrastructure. However, for a country that has been building cities relatively more densely concentrated than was desirable, there should also be room to examine the option of living comfortably at a lower density. The argument for making everything more compact begins from the assumption that building and maintaining all forms of infrastructure are the function of government. There is the example of Shimojyo village in Nagano prefecture where a reduction in the cost of infrastructure is being sought through the local government supplying materials and local residents building the roads. Naturally there are limits, but a more conscious effort is being made to explore how much maintenance of infrastructure is possible without relying completely on taxes.

■ Moving from localized optimum solutions to overall optimum solutions

We have so far discussed some of the fundamental problems that arise when considering sustainable society but I feel, with regard to the threats stated, that the latest German Machidukuri methods introduced in this book appear to approach all of these problems in a truly comprehensive way. German examples often go beyond the sphere of local government action, and are more wide-ranging than Japanese ones. It is not simply a question of land area. The approach itself is comprehensive and integrated. In Japan, the discussion is about localized optimum solutions such as housing estate optimum

テムを内包しているやり方なのだ。部分最適解の案に対して、なるべく平等に税金をばらまくことを是とする我が国が学ぶべきことは、この点にあろうと考える。

solutions, or what could be termed project optimum solutions. Of course, there are comparable cases at local government level in Japan such as the one in Toyama City, but these cases are overwhelmingly the minority. This is probably due to the Japanese approach to defining what a project is.

In most German cases, the set-up is that, with regard to a given social issue, a plan is carefully selected through a business project competition at a national or corresponding local level, and the approach selected then plays a leading role in resolving that social issue. An approach chosen in this way is thus brought to bear on the relevant issue in a wide context, and its success or failure make it a precedent to be taken into account when framing policy for the future. This is why it is required to be both comprehensive and integrated. It is not simply idea implementation. To put it another way, this is a method that encompasses a system which integrates a number of localized optimum solutions into an overall optimum solution. I believe that, as a country that approves of handing out tax money as fairly as possible between plans offering optimal local solutions, this is something from which we in Japan have much to learn.

目次

序文　サステナブル社会ー脅威からの解放の先にあるもの　大月敏雄（東京大学大学院教授）・・・・・・・・・・・・・・・14

第1章　サステナブル社会のまちづくりの本質とは 東京会議1・・・・・・・・・・・・・・・21

- 1-1、住み続ける住環境についてー可変性を備えた建築が重要　内田祥哉（東京大学名誉教授）　22
- 1-2、グローバル化とサステナブル社会のまちづくりー政策形成の立場から　井上俊之（（一財）ベターリビング理事長）　24
- 1-3、サステナブル社会のまちづくり研究についてー住環境資産の活用　澤田誠二（明治大学元教授）　27
- 1-4、日本のまちづくりシステムの課題と展望　大村謙二郎（筑波大学名誉教授）　30
- 1-5、サステナブル社会・ドイツのまちづくりーそのマネージメント　H. シュトレープ（ドイツ・プランナー）　32
- 1-6、討議とまとめ：プロジェクト・アプローチ、プランナー・コーディネーターの役割　36

第2章　サステナブル社会のまちづくりー日本とドイツにおけるプロジェクト 東京会議2・・・・・・・・・・・・・・・39

- 2-1、イノベーションシティ・ボトロップのまちづくり　G. ロエル（NRWジャパン社長）　40
- 2-2、IBA チューリンゲン 2013 – 2023 地域再生プロジェクト　H. シュトレープ（ドイツ・プランナー）　43
- 2-3、日本の団地再生・まちづくりの現状と課題　奥茂謙仁（市浦ハウジング＆プランニング）　48
- 2-4、「箱」の産業から「場」の産業へーまちづくり建設産業の転換　松村秀一（東京大学大学院教授）　53
- 2-5、ゲーラ2030まちづくりプロジェクトの仕組み　R. ミラー（ゲーラ市副市長）　56
- 2-6、富山市ー環境未来都市ーコンパクトシティ・コンセプト　中村圭勇（富山市環境未来都市推進係）　59
- 2-7、パッシブタウン黒部モデル　小玉祐一郎（建築家、神戸芸工大教授）　62
- 2-8、討議とまとめー日本の都市成長、コンパクトシティのプランニング、合意形成の実態　65

第3章　神奈川・横浜で取組む生活・産業・住環境の再構築 横浜会議・・・・・・・・・・・・・・・67

- 3-1、戦後70年の神奈川のまちづくりの変遷とこれからの取組み　猪股篤雄（神奈川住宅供給公社理事長）　69
- 3-2、汐見台団地の計画と成長の経緯、これからの課題　箆健夫（神奈川住宅供給公社専務理事）　72
- 3-3、神奈川における都市計画の発展と現状　鈴木伸治（横浜市立大学教授）　75
- 3-4、討議とまとめー現場を見てポジティブに考える、コンパクトシティの意味、国際交流のあり方　78

第4章　関西・滋賀県で考えるサステナブルなまちづくり 滋賀会議・・・・・・・・・・・・・・・81

- 4-1、滋賀のサステナブル社会のまちづくりのあり方　仁連孝昭（滋賀県立大学名誉教授）　82
- 4-2、ドイツのコンバージョン、リノベーションの建築デザイン　松岡拓公雄（建築家、滋賀県立大学教授）　85
- 4-3、討議とまとめーまちづくりプロジェクトの始め方、地域コミュニティに特有の課題　87

第5章　グローバル化するサステナブルなまちづくりの課題　澤田誠二（明治大学元教授）・・・・・・・・・・・・・・・89

（資料）
- 6-1、講師プロフィール　98
- 6-2、国際シンポジュウムの計画　澤田誠二　100
- 6-3、都市を耕すことー持続可能な生活・産業・文化　NJ. ハブラーケン　104
- 6-4、IBAエムシャーパークの実験　K. ガンザー　110　IBAエムシャーパーク方式の特徴　K. クンツマン　114

List of Contents

Preface   Sustainable Society-getting liberated from the threats and beyond   OTSUKI Toshio (Prof. Univ. of Tokyo) ···· **14**

Chapter 1   Sustainable Machidukuri – Basic principles Tokyo Conference1 ······························ **21**

  1-1 Long-life Building – Possibility of Adaptable Building   UCHIDA Yoshichika (Prof. Emer. Univ of Tokyo)   22

  1-2 Machidukuri Policy of Japan since 1980s INOUE Toshiyuki (the previous Director, Bureau of Housing, MLIT)   24

  1-3 Comparative Study on Sustainable Machidukuri in Japan and Germany   SAWADA Seiji (Meiji University))   27

  1-4 Japanese Machidukuri System and Current Issues   OMURA Kenjiro (Prof. emer. Tsukuba University)   30

  1-5 Sustainable Development Through Machidukuri in Germany   STRAEB Hermann (Town planner, Germany)   32

  1-6 Discussion and Conclusion – Project approaches, and roles of planners and coordinators   36

Chapter 2   Sustainable Machidukuri Projects in Japan and Germany Tokyo Conference2 ············ **39**

  2-1 Innovation City - Bottrop Machidukuri Project   ROEER Georg (NRW-Invest Japan)   40

  2-2 IBA Thüringen 2013 – 2023 Regional Redevelopment Project   STRAEB Hermann (Town planner, Germany)   43

  2-3 Recent Housing Refurbishment Projects in Japan   OKUMO Kenji (Ichiura Housing & Planning)   48

  2-4 From "Box"-making to "Place" -making industry – Change of Machidukuri in Japan   MATSUMURA Shuichi (Prof. Univ. of Tokyo)   53

  2-5 Gera 2030 Project – Integrated urban development project   Miller Ramon (Vice-Mayor of Gera City)   56

  2-6 Toyama City Environment Future City & Compact City   NAKAMURA Keiyu (Project manager of Toyama City)   59

  2-7 Passive Town Kurobe Model and Machidukuri   KODAMA Yuichiro (Architect, Prof., Kobe AC Univ.)   62

  2-8 Discussion and Conclusion – Urbanization, compact city concept and consensus building   65

Chapter 3   Development of Life, Industry and Housing in Kanagawa Yokohama Conference ············ **67**

  3-1 Machidukuri Development in Kanagawa Region and Future Strategy   INOMATA Atsuo (KHSC, Chairman)   69

  3-2 Shiomidai Housing Estate Project – Overview of its History and the next Programs   SHITOMI Takeo (KHSC. Managing Director)   72

  3-3 Town Planning and its Issues of Kanagawa Area   SUZUKI Nobuharu (Prof. Yokohama City Univ.)   75

  3-4 Discussion and Conclusion – Positive thinking on site, Compact City prospect, International exchange   78

Chapter 4   Sustainable Development of Shiga Area – Community and Environment Shiga Conference ····· **81**

  4-1 Community Development for Sustainable Shiga   NIREN Takaaki (VP, Univ. of Shiga Prefecture)   82

  4-2 Learning from Conversion and Renovation Design in Germany   MATSUOKA Takeo (Architect, Univ. of Shiga Prof.)   85

  4-3 Discussion and Conclusion   87

Chapter 5   Sustainable Machidukuri in Globalized Societies SAWADA Seiji (Meiji Univ. Prefecture) ·················· **89**

Reference Materials   6-1 Paneler's Profile Data   98

  6-2 Preparatory Study for the Symposium   100

  6-3 Cultivating Built-Environment – Sustainable Living, Industry and Culture   NJ. Habraken   104

  6-4 Nachhaltige Regionalentwicklung durch die IBA Emscher Park   K. Ganser   110   K. Kunzmann   114

第 1 章 サステナブル社会のまちづくりの本質とは
Chapter1 Sustainable Machidukuri - Basic principles

東京会議 1
Tokyo Conferense 1

内田祥哉（文責：内田）
Prof. UCHIDA

使い続けられる住環境について
－可変性を備えた建築が重要
Long-life Building –
Possibility of Adaptable Building

日本 / Japan
ドイツ / Germany

井上俊之（文責：井上）
Mr. INOUE

グローバル化とサステナブル社会のまちづくり
－政策形成の立場から
Machidukuri Policy of Japan since 1980s

澤田誠二（文責：前川）
Mr. SAWADA

サステナブル社会のまちづくり研究について
－住環境資産の活用
Comparative Study on Sustainable Machidukuri in Japan and Germany

大村謙二郎（文責：前川）
Prof. OMURA

日本のまちづくりシステムの課題と展望
Japanese Machidukuri System and Current Issues

ノルトライン・ヴェストファーレン州とチューリンゲン州
North-Rhine Westphalia (NRW) and Thüringea

H. シュトレープ（文責：前川）
Mr. STRAEB

サステナブル社会のドイツのまちづくり
－そのマネジメント
Sustainable Development through Machidukuri in Germany—
Management and Tools

1-1 使い続けられる住環境について―可変性を備えた建築が重要

Long-life Building – Possibility of Adaptable Building

内田 祥哉
UCHIDA, Yoshichika

私は4年前に「建築の寿命」について、ここでお話ししたことがあります。その中で古くから日本にあった特有の木造建築は、江戸時代になって建物としても建物づくりのシステムとしても完成されたことを述べ、それが"住まい方の変化に柔軟に対応できる可変性"を持ち、今の社会が求めている"使い続けられる建築"に一致するものであると述べました。

今各地で進められている団地再生プロジェクトの様子を聞くと、まずは住民合意の形成にコーディネーターの皆さんが苦労しているようです。この住民の意見には、60年前とは隔世の感があります。団地建設が始まった当時は「何が無くても住宅が欲しい」と誰もが考えていたのに対し、40年前には"一世帯当たり一住宅"の生活が達成されました。以後日本の社会は"豊かさ"を求めて走り続けました。社会全体が豊かになれば、一人ひとりの欲求も多様化し、高度成長の結果、モノあまりの時代になり、膨大な量の「空き家」をかかえる社会になっていきます。

現在進められる団地の再開発のプロジェクトでは、事業に伴うコストとリスクの分担について住民の合意が必要になります。この分担には住民の負担と事業者側の負担とがあります。その内住民の負担だけを見ても、世帯ごとの費用には、リニューアルした後のライフスタイルの変化も予想されるので、住人全体の一致にたどり着かないのです。

そこで、住宅という"生活の器"の側に可変性・順応性を持たせて、個人の欲求や世帯ごとの多様な価値観に合せられる"使い続けられる建築"に関心が向くというわけです。

木造建築は木材を山から取りますから、山を育むことも木造建築システムを構成する重要な部分です。また資材の流通、加工と現場での組立てを経なければ木造建築はできません。出来上がった建物についても、清掃などの日常の手入れ、修理、間取りの変更、増築、更には建て直しなどが必要になるので、そのための技術と職人とが揃って"使い続けられる建築"が本当に"使い続けられる"のです。

住宅不足時代の日本は住宅団地による生産の集約化により生産性の向上を図りました。急速な都市化の進む中で、団地住宅のほとんどを壁構造の一体式コンクリートとしたために、年を経て、建物の改造が不自由です。住宅部品の開発やら工事の進め方について、伝統の木造システムの在り方から大いに学ぶ必要があります。

サステナブル社会のまちづくりの核になるのは住宅建築です。住宅建築のあり方については、以上のように総合的に見ないといけないのです。皆さんの議論が「旧来の陋習を打ち破り」グローバルな提案となることを期待します。

Four years ago in this very spot, I talked about the history of architecture, and in my speech I noted that the wooden houses typical of Japan from the beginning of our history was completed as a system for building houses during the Edo period; and that its flexibility allows it to be adapted to people's lifestyles. This truly represents the sustainable building that is required by present society.

Today as we listen to the many ideas that people express about the various housing rejuvenation projects, I feel that, first of all, there has to be consensus between residents and the coordinators that is sometimes difficult to reach. The perspective regarding residents has changed significantly since the 1960s. In the '60s when the large housing estates were built, most people said that having a house was more import that anything else. Japan soon achieved its goal of "one house, one family" at the beginning of the '70s. And even after attaining that goal, Japanese society continued to pursue affluence. When affluence became widespread, people's desires diversified, which resulted in high economic growth that led to a flood of products. Critical situations emerged, such as an excess in vacant housing that we now have in our midst.

There has to be consensus between residents and builders on sharing costs and risks when houses are renovated. These burdens borne by residents and builders can be changed after building renewal and renovation. After hearing this difficulty faced by residents and builders, I would suggest looking into the "adaptability of the traditional wooden house system." I believe that housing planning will be liberated by introducing this concept and a meaningful agreement reached between residents and builders as well as the housing management organization.

To construct houses of wood, we can acquire timber from Japan's forests. Therefore we need to cultivate forests, which is crucial if the system of constructing wooden houses is to work. In addition, we need a way to distribute the timber, as well as skilled workers capable of assembling the components on the building site. Without the production team, houses cannot be constructed or maintained for a long period of time. This Japanese system of constructing wooden houses had developed from its earliest beginnings until its completion in the Edo era, a period of almost 1,300 years.

This tradition, the technology as well as organization of the industry, was passed along even when modern building materials and technologies were adopted in the urban development process during the modernization of Japanese society. The inherited technologies and skilled workers are still very much a part of today's society.

As for the production of housing estates, in response to the rapid economic development, most apartment houses were built on concrete structures which embedded the subsystems, such as windows, doors, kitchens, and bathroom, that cannot be replaced when their use times are over. The building industry sector must now tackle this problem, since the volume of housing supplied since '60s is huge, making this a very serious problem. If we could look closely at that traditional Japanese system at that time of the mass housing supply, we could learn a lot from it and be well poised to solve current problems.

As we meet today to consider appropriate ways to plan, design, construct, and maintain a living environment for sustainable community, I would suggest that our traditional system offers us good ideas, and I hope that all the people from the different sectors will discuss them, leaving behind the usual banalities, and address the central issues of "how the houses are to be built and maintained."

Finally, I hope that the discussions and their results will contribute to the sustainable development in Asia and other regions of the world.

図1 伝統木造住宅の造られ方（仕組み）
Fig. 1 Building system of traditional Japanese house

図2 大阪ガスの実験住宅 NEXT21
Fig.2 Osaka Gas's experiment building, NEXT 21, in Osaka
出典: 大阪ガス

1-2　グローバル化とサステナブル社会のまちづくり―政策形成の立場から
Machizukuri Policy of Japan since 1980s

井上 俊之
INOUE, Toshiyuki

■ 我が国の国土・人口の状況

まずは日本の国土と人口の状況を見ていく。図1は主要国の可住地人口密度、人口を表したものだが、日本は1,115人/km²。先進国のなかでも可住地の人口密度はきわめて高い。ざっとドイツの3倍の密度である。考え方によってはコンパクトな国土の構造ともいえるのではないか。

次に人口動態を見ると、周知のように日本の総人口は2010年をピークに減少し、高齢者人口が増加する一方、生産年齢人口と若年人口は減少している。

都市圏と地方圏別に分けて、1980年以降の人口推移をみると、都市に人口が集中し、逆に非都市圏の人口は30年前から下がり続けていることがわかる。三大都市圏と政令指定都市圏では人口ピークはこれからだが、地方都市圏ではピークを終えた。（図2）

また、市街地の面積は人口以上に拡大してきた。三大都市圏を除く県庁所在都市の人口集中地区[DID]の面積は50年間（1960年～2010年）で3.6倍に広がっているものの、この間の人口増加は1.6倍である。都市部に人口は集中しているけれども市街地は拡散しており、密度は薄くなっている。

また、社会問題化している空き家については、日本の住宅戸数6,063万戸のうち853万戸にのぼる（2013年）調査の速報値。長期不在・取り壊し予定の空き家は構造的にこれからもまだまだ増える見通しである。（図3）

■ 住宅・建築物のエネルギー利用の状況

日本のエネルギー消費の現状と推移を見る。住宅・建築物部門が全エネルギー消費の1/3を占める。かつては1/4だった。他部門と異なり増加が著しい。（図4）

運輸部門、産業部門では設備・車両の省エネルギー化や事業者の努力で消費量の削減に成功してきた。

また、産業部門については海外移転の影響もある。住宅部門については、温水洗浄便座やパソコンなど電子機器の普及があるうえ、単身世帯など世帯数が増えていることも消費量増加の一因である。

また、建築部門は、オフィスや店舗等の床面積が増えていること、そしてコンビニエンス・ストアや飲食店の24時間営業など使用時間が増えていることが消費量を押し上げている。

我が国のエネルギーの使い方は他の先進国との比較ではどうなのか。図5は世帯当たりのエネルギー消費量を国別に記したもの。図の赤い部分が暖房に関する消費エネルギーだが、ひと目見てわかるように日本は暖房の消費量が欧米に比べて極めて少ない。その反面、給湯、照明・家電の消費量が（アメリカほどではないが）多い。風呂を使う習慣があるため給湯が占める割合が高い。

ドイツでは『断熱改修をすれば数年で元手がとれる』といわれているが、日本の場合はそもそも暖房の消費量が低いため、同じようにはいかない。ただし、国土が南北に広がる日本では地域差も無視できない。暖房消費量が高い北海道・東北と、暖房がほぼゼロの沖縄を一括りにはできない。国土の多様性も考えるべきである。

■ Housing situation of Japan--population and urbanization

1. The density of 1,115 people per 1 km² inhabitable land indicates that Japan is at the top among the countries (Fig.1). It is almost three times that of Germany. Japan could be a "compact" country.
2. Although the general population has been in decline since 2010, the aged population is increasing, as is the number of both "productive people" and "young people" (Fig.2).
3. The population change with respect to urbanized and non-urbanized regions since 1980 indicates that the concentration of people in the urban areas is still taking place and that the decline in the surrounding regions had already begun 30 years ago. The population will peak in the three metropolises of Tokyo, Osaka, and Nagoya, as well as in the Seirei-Shitei Toshi (Cities) regions. The population of local cities has already peaked.
4. The area of urban districts has increased more rapidly than the population. While the area of the DID (Dense Inhabited District) had increased 3.6 times from 1960 to 2010, the population of the areas had only increased 1.6 times. This means that the population in the areas is spread out at a lower density.
5. The problem of "vacant housing" is one of the important social issues in need of public solutions (Fig.3).

■ Energy consumption in housing & building sector

1. Energy consumption in the housing and building sector is now about one-third of total energy consumption in Japan, having previously been one-fourth (Fig. 4).
2. Energy savings in the transportation and industry sectors have successfully been achieved through changes in the building service systems and by improving vehicle designs. The decrease in energy consumption in the industry sector has been influenced by Japanese industries relocating overseas.
3. The energy consumption increases in the housing sector reflects the rapid popularization of home electric appliances and electronics.
The increase of "single households" in the younger generation as well as among the elderly has been significant over the past 20 to 30 years.
4. Japan's specific energy usages can be seen in Figure 5. The energy used for warming is very little compared to other countries. However, that consumed for hot water, lighting, and home appliances is greater in Japanese homes. The amount of energy used for bathing is also significant.
5. Japanese measures to save energy are very much influenced by these lifestyles and climate conditions. The measures take large regional differences into account, since the country, although small, is heterogeneous in nature, both geographically and culturally.

■ Energy saving efforts in housing and building sector

1. Subsequent to the "Act on the Rational Use of Energy, 1980," the following legal measures were introduced:
　--Regulations and Incentives for Newly Constructed Housing
-Housing Quality Assurance Act
--Approval of Long-life Quality Housing
--CASBEE (Comprehensive Assessment of Building Environ- mental Efficiency)
--Building Energy-efficient Labeling System, etc.

■住宅・建築物の省エネルギー化の取り組み

住宅・建築物の省エネルギー化に向けた国の取り組みを振り返ると、第一に1980年制定の省エネルギー法に基づく建築・住宅新築時の規制誘導措置がある。省エネルギー基準を定め建築主に努力義務や省エネルギー計画の届け出義務を課すものだが、基準の強化や届け出対象の小規模化を順次実施して、当初に比べると順次強化されてきている。第二に性能表示・情報提供等の取り組みである。

住宅品確法による住宅性能表示制度、長期優良住宅認定制度、建築環境総合性能評価システム（CASBEE）、省エネルギー法に基づく「住宅省エネラベル」などである。

さらに最近では建築物省エネルギー性能表示制度（BELS）も始めたところである。第三に助成策である。住宅・建築物省CO_2先導事業、住宅エコポイントなどの補助や住宅金融支援機構のローン金利優遇・減税などの優遇措置により、省エネルギー化を進めてきた。

■コンパクトでサステナブルなまちづくり

個々の住宅・建築ではなく、面的なまちづくりではどうなのか。コンパクトなまちづくりに向けた法律について振り返ってみる。まず、コンパクトシティとは銘打っていないものの2006年に行なった「まちづくり3法」の改正がこれにあたるだろう。大規模小売店舗立地法、中心市街地活性化法、都市計画法（ゾーニング規制）の改正により、大規模集客施設の立地規制、中心市街地活性化への支援措置の強化等により商店街の空洞化を防ぐことを目的としたものである。

2012年には、東日本大震災によるエネルギー危機を契機に都市の低炭素化の促進に関する法律が成立した。地方公共団体が低炭素まちづくり計画をつくり、建築・住宅や公共交通などの低炭素化、下水道の潜熱利用などにより、低炭素化とエネルギー利用の合理化の取り組みを促すものである。

2014年には、コンパクトシティを正面から取り上げた都市再生特別措置法の改正（コンパクトシティ法）を行った。大都市圏を含む都市の人口減少・高齢化の進行に対応し、まちをコンパクトに作り直していこうというものである。立地適正化計画の策定とそれに基く都市機能や居住の誘導区域の設定が柱である。また、財政、金融、税制や容積率特例によるインセンティブの付与でコンパクトな構造への転換を図る。まだ、できたばかりで運用はこれからである。

■環境モデル都市・環境未来都市について

温室効果ガス排出の大幅な削減など低炭素社会の実現に向け、高い目標を掲げて先駆的にチャレンジする都市として富山市を含む23都市を環境モデル都市として選定している。また、環境・超高齢化に対応した人間中心の新たな価値を創造する都市をコンセプトとして、11都市・地域を環境未来都市として選定している。先進的都市の実践的取り組みを広めていくことが重要である。

■今後に向けて

住宅・建築部門の省エネルギーを一層進めることが必要なことは言うまでもない。政府は新たなエネルギー基本計画を2014年4月11日に発表したが、財産権に配慮しながらも、2020年までに新築住宅・建築物について段階的に省エネルギー基準への適合を義務化するとしている。そこでカギになるのは、一戸建て住宅の半数を担う大工と工務店だ。この方々が省エネ基準を理解し、守っていただくことがポイントになる。大規模な建築物から義務化することとなるが、現在立法の準備を進めていると聞いている。

図1　各国の人口と可住地人口密度
Fig. 1 Population and desity (per inhabitable land) of the countries

国名	可住地人口密度（人／km²）	人口（億人）	
日本	1115	1.28	Japan
アメリカ	52	3.14	USA
イギリス	299	0.64	UK
ドイツ	341	0.80	Germany
フランス	171	0.64	France
中国	187	13.61	China
シンガポール	8223	0.05	Singapore

図3　急増する空き家
Fig. 3 Rapid increase of vacant housing (1948 – 2008) in Japan
■ Number of houses in stock
■ Number of households
Total not-lived houses: 8,530 mill in 2008

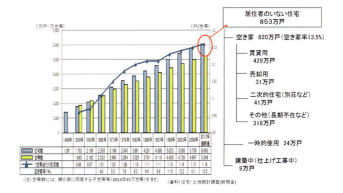

図4　日本のエネルギー消費の推移と現状
Fig. 4 Energy consumption in Japan
■ Transportation　■ Housing sector　■ Industry sector

図5　主要国の世帯当たりのエネルギー消費量（GL/世帯・年）
Fig. 5 Energy consumption per a household of the countries:

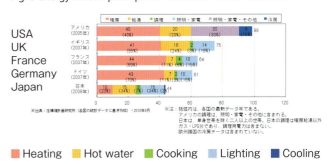

■ Heating　■ Hot water　■ Cooking　■ Lighting　■ Cooling

これまでの日本の政策は、人口が増えていく、経済規模も拡大していくという右肩上がりで想定していた。しかし、人口減少・縮小を受け入れる政策への転換がサステナブルな社会の実現につながる。例えば空き家問題。老朽化で倒壊の恐れがある「危険空き家」の除去に対する「空家等対策の推進に関する特別措置法」が議員立法で成立(2014年11月19日)している。

しかし、危ない空き家は増える一方であるが、これに対する対策はいわば対症療法に過ぎない。空き家の利活用・適正管理、危険化する前の除却の推進により危険空き家化の防止を図ること、さらにその手前で都市の集約化、コンパクトなまちづくりを進めまちづくりの観点からの空き家発生の防止を図ることが重要である。住宅政策の観点からは良質な住宅、建替えへの誘導を図り空き家の潜在的予備軍を増加させないようにすることも必要だろう。

都市政策についても現在の制度的枠組みの多くは、人口増とそれに伴う、開発圧力のコントロールが課題であった時代に構築されたものである。今後はその延長線上ではない新たな施策体系を整備し、提示していくことが必要である。コンパクトシティ法はまさにまちづくり政策を転換させるものである。何十年もかかって広がった都市を数年で狭めていくのは無理だ。息の長い取り組みが必要で、問題を先取りする政策は非常に困難だが、やっていかなければいけない。のちほど紹介のある富山市の政策は、市長がリーダーシップをとって郊外部の理解も得ながら時間をかけて進めたものである。法律としては今回大きく舵をとったので今後は富山市のような取り組みの蓄積を通じて制度を充実させていくことも必要だ。

課題は非常に大きいが、今日のシンポジュウムはその解決の一助になるであろう。

■ Measures for Compact and Sustainable Machidukuri

1.The measures, which go beyond houses/buildings to streets/districts, initiated the improvements of three Machidukuri Acts in 2006:
   --Act on the Measures by Large-Scale Retail Stores for Preservation of Living Environment
   --Chushin-Shigaichi (Downtown) Revitalization Act
   --Town Planning Act
2. The Low-Carbon City Act was introduced for the Local Public Body in 2012, to plan low-carbon cities and infrastructure.
3. This year, 2014, the so-called "Compact City Act" was adopted, as was the "Act on Special Measures Concerning Urban Renaissance," the purpose of which is to reorganize urban spaces and buildings in accordance with the appropriate planning of urban functions and the targeted area.

■ "Eco-Model City" and "Future City"

The new measures were enacted to encourage cities making efforts toward sustainable society in Japan. The government nominated:
1. Eco-Model Cities, because of their efforts to change to a low-carbon community by, for instance, significantly reducing greenhouse gases, etc., and achieving their goals by using advanced technologies (e.g., LRT in Toyama).   23 cities nominated, including Toyama.
2. Future Cities because of their efforts to adapt to environment-friendly and rapidly aging communities. 11 communities.

■ Perspectives

1. The New Basic Act on Energy, also enacted this year, requires activities of energy producers and consumers to comply with the standards.
2. The Act on Special Measures concerning the Promotion of Measures for Vacant Buildings, enforced also this year.
The government has steered its way toward sustainable society. The efforts of Toyama, for example, should be applied to others cities.

図2　都市圏・地方県別の人口の推移
Fig. 2 Population changes in different urbanized areas in Japan　● 3 Metropolises　■ Seirei-Shitei Cities　△ Other Cities

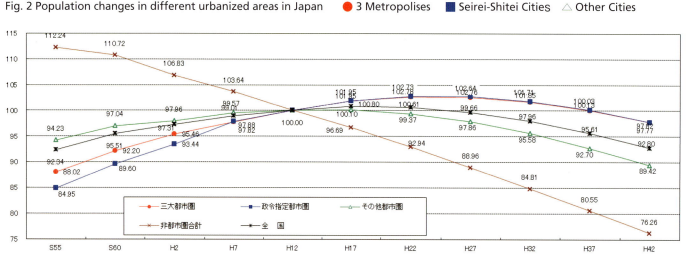

1-3 サステナブル社会のまちづくり研究について―住環境遺産の活用
Comparative Study on Sustainable Machidukuri in Japan and Germany

澤田 誠二
SAWADA, Seiji

■対比から何が得られるか

本シンポジュウムの企画立案者である澤田誠二氏は、今回の開催主旨を含めて「サステナブル社会のまちづくり研究」について踏まえておくべきポイントを明らかにした。

世界的に見ると、産業革命以降は第一次世界大戦、第二次世界大戦という大きな出来事を超えて、まちづくりや住宅づくりは進歩してきたが、「その過程であとあと十分に使っていけるような正しい遺産をつくってきたかというとそうではない」と言う。

その認識をもとに戦後、つまり20世紀の中ごろから都市計画・まちづくり（建物を計画してデザインし、構築して運営管理していく）は発展してきたが、今その形態と結果を照らし合わせることが、これからのサステナブル社会のまちづくりに役立つと考えている。「日本とドイツを対比的にとらえることで、この回答がどこまで得られるのかを試すこと」が今回のシンポジウムの主旨なのだ。

ポイントは5つある。（1）現代のまちづくりシステムは1920年代に形成されたという前提で、日本とドイツ、あるいはヨーロッパ圏や他の文化圏まで考えること、（2）国際化が進み、社会・経済が発展し、都市化まちづくりが進んだこと、（3）都市とまちのつくられ方を振り返る必要があること、（4）今までのまちづくりシステムが形成したものをカテゴリーに分け、なおかつ日本とヨーロッパの対比から考えること、（5）サステナブル社会のまちづくりシステムのイノベーションのポイント――である。順を追って説明しよう。

■都市化やハウジングを見直す

現代のまちづくりシステムは1920年代に形成されたが、ポイントは、それまでの自助的かつ自主的システムを母体として、社会の近代化に対応するモダンなシステムへと進化し、それが産業開発・都市開発に用いられた点にある。「まちづくり」の産業化のための新材料・技術は、生産性向上ツールの規格化と法制化によって進んだが、さらに大事なことは「まちづくりにかかわる面で人の専門資格化がはじまったこと」と語る。新しい材料・技術を開発するだけでなく、それを用いる人を育てなければうまくいかないからだ。

国際化によって社会・経済が発展し、都市・まちづくりが進んだことは日本の戦後を振り返るとよくわかる。都市空間の高度利用と建築設備の高度化が車の両輪だった。「40年前と現在の工事費の見積を見ると建築設備の費用が全然違う」と指摘する。さらにクルマ、鉄道、航空機などモビリティシステムの発展に加え、多量生産・供給システム（マスハウジング）の部品化と品質管理を厳格にすることで生産性のよいシステムを構築してきたのだ。

社会成長の終息期で、価値観が多様化しはじめた1970年代以降で重要なのは「多様なニーズに対応できる産業システム」だ。都市化やハウジングをあらためて見直すなかで、私たちは都市というものを何のためにつくってきたのか、これからは何のためにつくっていくのか、そのためには都市をどう運営していけばよいのかという3つの視点が、サステナブル社会のまちのあり方を考えるうえで意識しなければならないことである。

■ Machidukuri system was formed around 1920 years and applied for until the high economic growth era

a. Based on traditional "self-sufficient, local" systems, the Machidukuri systems was developed for the "modern societies".
b. Evolution in 1920 years in terms of building materials and technologies and also internationalization made the system of higher efficiency, quality management and market competitiveness. System's tools: standard, product specification and profession qualification.
c. During post-WWI and –WWII times, the systems were mproved to adapt to rapid and efficient urban development.
d. The new technologies; industrialized production, mass housing supply system, prefabrication were adopted.
e. Along with these progresses in the housing field, the distance between the planners and "planned side" became larger . Which is quite different from "self-sufficient" system.
Some new concepts had been proposed to make housing more adaptable to variety of living needs, such as " Open Building concept." (Fig. 2)

図1 ロンドン、パリ、ウィーン、ベルリン、東京の人口変化
Fig.1 Population development: 1750~1990,

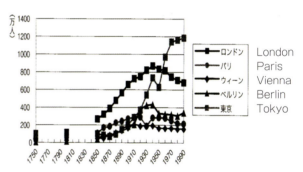

図2 オープンビルディングの考え方―レベルごとの意思決定
Fig.2 Open Building model: Living environment levels vs Client Urban Tissue level / Base Building level / Fit out level

■ Overview of results from Machidukuri activities and the objects in sustainable society

a. Results in different areas: Urban areas, Housing estates, Infrastructure, Suburban areas, and Industrial areas

b. Each result regards to "artificial environment asset" of different characters, and different potentials in sustainable society.

■カテゴリーごとに分けて考える

では、あらためて都市とまちのつくられ方を考えてみよう。かつての集落が「まち・都市」に発展してきた。まち・都市は産業の拠点であり、通商都市の性質をもつ。さらに文化活動の場でもある。

ここで澤田氏は、ロンドン、パリ、ウィーン、ベルリン、東京の人口増加の経年変化を見せた。東京だけが急激に人口が増え、今も他の4都市をはるかに上回る人々が住んでいる。（図1）

続いてパリ、ベルリン、東京の航空写真を投影した。「日本の現代都市（東京）はヨーロッパと変わらないことがわかる。さらに、日本の都市の上に堆積している古い技術や材料の量は、短い間に堆積したものが多いことがわかる。建物のデザインもそうだ」と話し、これからの都市のあり方をおおづかみで考えるときのヒントになるだろうと述べた。

今までのまちづくりシステムが形成したものとして、「都心部」「ハウジング」「都市インフラ」「郊外」「産業地帯」と5つのカテゴリーに整理した。（表1）

高層化・高密化、多機能化建築の建設、市街化エリアの拡大が進む都市の中心部では、「コンパクトシティ化」や「リアルな環境の形成」が求められている。ハウジングは、都市周辺に開発された大規模団地では老朽化や高齢化、空き家などの問題を抱える。コミュニティ、住環境、ハウジング経営をそれぞれ再生しなければならない。

都市インフラは多量・高速輸送が人口集中を促進したが、これからは子どもやお年寄り、障がいのある人たちも利用しやすいようなエリア内の交通システムの充実が求められている。

郊外については、井上氏が基調講演で触れていたように市街地の拡大（拡散）が進んでいる。それが農業用地や緑地を圧迫し、生活の場としての質が落ちている。「国土計画レベルの見直しが必要だ」と考えている。

産業地帯では「負の遺産の総合的再生」が望まれる。一例としてドイツの「IBAエムシャーパーク」を挙げた。ルール工業地帯など重厚長大型の工業地帯は1980年代に「20世紀社会のつくった負の遺産だ」と呼ばれるようになる。その再生に取り組み、800km²もの広大な地域全体で「緑と水のネットワーク」を30年がかりで再生したものだ。（IBAエムシャーパークについては後ほど詳細な報告がある）

■日独の成果を対比的かつ具体的に

最後にサステナブル社会のまちづくりのために必要なシステムのイノベーションのポイントを5つ挙げた。

まずは日本とドイツを比較するという観点。人種も社会システムも違うが、何が違うのか、どの部分を比較すると有意義かを整理する必要がある。

「ただし、サステナブルコミュニティと住環境の基本的なコンセプトは共通したものがあるかもしれない」と言う。人が住んでいる建築環境をレベルごとに見て、それぞれの耐用年数や環境を誰がどのように決めるのかまで含めて考えると世界共通なものがあるかもしれない。「それがオープンビルディングというもので、私は40年前から研究している」と述べた。

オープンビルディングとはオランダの建築家、ニコラス・ジョン・ハブラーキン氏が40年ほど前に提唱したもので、「インフィル」「スケルトン」「アーバンティッシュ」という3つの住環境レベルに分けて考えるもの。インフィルは住宅で、住宅が納まる建物がスケルトン。そして建物（スケルトン）が集まった団地は「アーバンティッシュ」という1つの都市構造となる。（図2、資料6-3を参照）

インフィルは居住者が決めるもので、改修までは10〜20年くらい。スケルトンは建築家や設計士が決めるもので、耐用年数は50〜100年く

Tab.1 Environment assets made by modern town planning and the possible use of them in sustainable society

Urban areas:	high use of urban space -- density, high-rise, and multiple functions high potential for investors / developers
Housing estates:	large housing estates around big cities strongly influenced by population change re-planning needs community consensus
Infrastructures:	high speed, big capacity rail and road Network to connect big cities to improve inner city mobility by LRT etc.
Suburban areas:	expansion of urban area and invasion of natural environment, and/or firming land land use control through eco-balance rules
Industry areas:	heavy industries infringed natural land while supporting the life of the people nature conservation and reuse of the asset

表1 近代都市計画の残したもの、サステナブル社会における使い方

都市中心部	都市空間の高密・構想化、多機能化 サステナブル社会でも高度な投資、開発の場
ハウジング	都市周辺に住宅団地を開発、社会発展の受け皿 再生は住民合意が必要
都市インフラ	クルマ、鉄道輸送を実現、産業・生活発展を支持 今後は都市内の高機能インフラ整備が必要
都市の郊外地	都市化が郊外地を侵害、農地利用を圧迫 自然環境・人工環境・遊休環境資産の管理が必要
産業地帯	重厚長大産業地帯を形成、経済発展を促進 "負の環境資産"と捉えて再生を計画する

■ Innovation of Machidukuri system for sustainable society

Projects in sustainable society are in a great variety.
Five points are needed for the improvement of present system;
a. Project creation in closer contact to the clients / users
b. Project process able to be changed when conditions altered
   i.e. PCM (project cycle management) (*).
c. Realization of project / building through competitive selection
d. Project resource provision; technical data and product data
   and other resource information should be provided so that
   the optimization works are made easily.
About PCM approach
Basic hypotheses;
a. A project is not a linear process
b. Urban development is implemented by many different stakeholders
c. Participation helps to improve quality, acceptance and sustainability (from Meiji Univ. seminar, 2011)

表2 サステナブルなまちづくりシステムに必要な条件

プロジェクトの創出	多様なプロジェクトに取組み、 高いクオリティの成果を出すこと
プロジェクトの運営	多数の利害関係者の意見を引出して 変化するプロセスを運営する必要あり
プロジェクトの実現	プロジェクトのクオリティを最適化 出来る工事者を選定する
プロジェクトの環境	多様なプロジェクトに、必要なリソース （技術、製品情報など）の提供

らい（現在は 200年くらい保つ技術もある）。アーバンティッシュは自治体が決めることで、耐用年数は200〜300年くらいと考える。（図2）

次にイノベーション志向のプロジェクトだが、課題としては住環境長寿命化、エネルギー、生活支援などがあり、それを解決するプロジェクトプロセスとして合意形成、意思決定、実施計画がある。

さらに、こうしたイノベーション活動をマネージメントするためには、政策の連携（省庁間の連携）、公的促進策（一般の助成とイノベーション促進策）、技術ノウハウ提供（オープンソース化）、連携体制づくり（連携ルールの明確化）を挙げている。（図4）

「日本とドイツのいろいろなまちの事例を取り上げて、具体的にどんなプロジェクトマネージメントが行なわれていて、どのような成果を上げているかを、対比的かつ具体的にとらえたい」と述べた。

参考（明治大学「サステナブル社会のまちづくり」セミナーより）
サステナブルなまちづくりに関する基本認識:
下記の状況により、プロジェクトは多様化する。
ー技術、産業、経済の複合化が進む
ー社会問題の多様化・複雑化が進む
ー政治のコンプレクシティ（複雑性）が拡大する
プロジェクトに取組む際に心得るべき3事項:
ープロジェクトは直線的には進まない
ーまちづくりには様々な利害関係者が参加する
ーそれら参加者の意見を上手く調整して、
クオリティを実現する

■江戸時代から明治時代そして現在までの東京の変遷 / Change of Tokyo: from 17th century until today

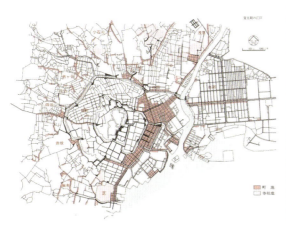
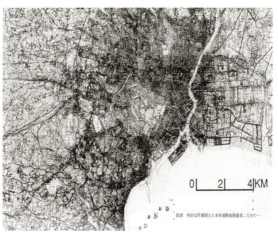

1-4　日本のまちづくりシステムの課題と展望
Japanese Machidukuri System and Current Issues

大村 謙二郎
OMURA, Kenjiro

■日本の都市計画はパッチワーク型

　都市計画を専門とする大村謙二郎氏は、このあとに登壇するシュトレープ氏との対比も踏まえて、日本の都市計画（まちづくり）システムの特徴や課題、展望について言及した。

　まずは「近代都市計画としての日本都市計画」の特質から。日本の都市計画は 20世紀初頭に現行の都市政策の原型ができたが「ヨーロッパやアメリカでも都市計画は 19世紀末から20世紀にかけてできたので、日本はそれほど遅れたわけではない」と言う。そして、日本の特性として、他国でどういうことをやっているかを旺盛に学ぶ傾向があることを挙げた。たとえばゾーニングはアメリカの導入システムを学んだものだし、都市区画整備はドイツの「アディケス法」（フランクフルト市長のフランツ・アディケスが1902年に制度化）を学び再開発の手法に役立てた。開発許可制度はイギリスの制度をもとにしている。1980年の地区計画制度はドイツの制度から学んでいる。

　「日本は、日本の特性、風土に対応した独自の都市計画をつくりあげるためのハイブリッド化を進めてきたけれど、逆の意味では各国の都市計画制度を寄せ集めてつくりあげたパッチワーク的な面があることも否めない」と指摘した。

　また、近代都市計画としての日本の都市計画には、次のような特徴があるという。
（1）成長、開発指向型都市計画
（2）都市更新・開発の高速度
（3）開発自由、建築自由が原則（強い土地所有権）
（4）土地利用コントロールの弱さ
（5）分権／参加が大きな課題

　日本の都市計画は、ひじょうに旺盛な成長・開発志向型だったうえ、急激に人口が増えたため、開発・都市更新の速度が速い。「これは今も変わっていない。このままでいくのか？」と疑義を抱く。土地所有権が強いため、土地利用コントロールの弱さにもつながっている。また、課題として「都市計画の権限を基礎自治体である市区町村に委ねていく分権化の流れ、計画策定、実施に住民を含め多くの主体がかかわっていく参加型まちづくりも推進されているが、まだ多くの課題を抱えている」と挙げている。

■高度成長期以降の都市計画

　次に、高度成長期から時系列で日本の都市計画の流れを追った。日本の現行都市計画の仕組みは1968年に制定された都市計画法によるが、ドイツでも1960年に現行の都市計画システムの原型ができあがった。

　日本の都市計画システムの特色は「土地利用」「市街地整備」「都市施設」という3つの要素から成り立っている。土地利用コントロールを行なう都市計画は、ドイツでは全国土を対象にしているが、日本では国土の40％くらいしかカバーできていない。また、市街地整備と都市施設では、急速な都市化に対応するため、区画整理、道路や橋梁、上下水道といったインフラ整備に重点が置かれていたと分析する。

　1950年代後半から1970年代前半、日本は高度成長が続いて、大都市圏に人口が集中し、大規模団地が各地につくられた。それだけでは足りなくて小規模住宅地開発がスプロール状に展開されるなど、旺盛な宅地需要、産業用地需要に対応するために、大量住宅供給、画一的・

1:Introduction: Japanese urban planning as a plot type of modern urban planning
　Development process of planning in the era of high economic growth
　-Stablishment of present planning system by 1968 law
　-Main contents of planning system and its characteristic: land use, urban development project, urban infrastructure
　-Background of planning in the era of high economic growth and its remarkable points:
rapid urbanization, esp. in three mega urban region, parallel emergence of sprawled development and planned housing settlement, huge space demand for housing and industry, mass production and mass supply of housing, uniform and standardized urban development

2:Urban planning from the age of post high economic growth: development process from 1970s
　• Change of surrounding of planning: oil crisis in 70s, environmental pollution, concern for natural environment, ecology, awareness of limit of growth,
-New trend of planning: concern for neighborhood environment, community based planning and participation for planning, interest for historical urban environment, conservation planning, district planning system of 1980, Machidukuri movement

3:Problems and policy of the era of bubble economy in the late 1980s:
　• New urban housing problems in inner area, new housing policy:
　• Master plan for housing, structural change of industry and logistics, progress of urban redevelopment project, conversion of industry site, financial crisis of public sector, Public Private Partnership, privatization

4:Trend of planning from 90s, urban development & housing projects from post bubble economy:
　• Influence of globalization, similar urban regeneration big project through PPP, PFI etc.,
　• Worldwide competition of urban development, utilization of world famous architect and urban designer for urban development,

5:Change of conditions and new trend of urban & housing development
　• Change of socio-economic conditions: demographic change, less kids and aging society, structural change of industry,
New demand of urban infrastructure, worldwide competition of mega cities,
　• wo vectors of globalization and localism, climate change, mega-hazard, renewable energy etc.
　-Regeneration, reuse & renovation of existing stock, priority for inner development instead of suburban development, compact urban and regional structure
　• Various key words for urban development & machidukuri (community based development): creative city, smart city, compact city, resilience, sustainable development, low carbon city
　• Various type of planning incentive

機能的都市開発プロジェクトの推進などがこの時期の課題であった。

1970年代以降のポスト成長期は、2度にわたる石油危機、さらに公害の頻出からの環境問題などによって、自然環境や生態系が再認識され、省資源、省エネルギーが重視されるようになる。また、身近な居住環境への関心が高まり、コミュニティ主体の計画や参加計画が策定されるようになった。

「歴史的市街地を大事にしよう、あるいは、地区レベルの計画システムを充実させていこうというまちづくり運動が徐々に脚光を浴びはじめた」と言う。

■求められるインセンティブ型計画

1980年代後半のバブル期は、急速に都市計画が進んだ。日本の都心では住宅が足りないという問題が発生。さらに、製造拠点の海外移転が相次ぐという産業構造の変化、それによって現れた遊休地で都市開発を進めていった。しかし、公共部門の財政危機が現出し、民間活力を活かした都市開発をいかに行なうかが課題となっていく。

バブルがはじけた後のポストバブル時代の都市計画は、グローバリゼーションの影響を受ける。また、PPPやPFIなどによる都市再生大規模プロジェクトが見られるようになる。あるいは海外との都市間競争が激しくなり、世界的な建築家や都市デザイナーを登用する動きも出始めた。

その一方、1995年1月17日に発生した阪神・淡路大震災により、被災地の救援に駆けつけた市民による草の根型のボランティア活動が広がり、その後のNPOの浸透につながる。いわゆる「ボランティア元年」である。これもまた都市計画の新たな潮流だと指摘した。

こうした状況の変化と都市・住宅開発の新たな動向について、「日本が直面しているのは、人口構造の大きな変化だ」と見ている。「世界でもっとも早く少子化・超高齢社会を迎えたフロントランナーの日本が、この問題にどう対応するかは海外からも注目されている」と述べた。

また、世界都市間競争やグローバリゼーションとローカリズムの2つのベクトルへの対応、気候変動および巨大災害への対応、東日本大震災を契機とした再生エネルギーへの取り組みが新たなカギになると予測。

一方、既存ストックの再生と利活用の重要性も説いた。「郊外開発よりも既成市街地の再生が優先されるべきで、集約型都市・地域構造に緩やかに移行することが課題」と語る

また、これからの都市開発・まちづくりに対するキーワード「創造都市、スマート都市、コンパクト都市、強靭な都市、持続可能な開発、低炭素都市、エコまちづくり」を挙げた。これらを1つひとつ意味あるものにしていくことが大切と述べるとともに、インセンティブ型計画の多様化が重要だと主張。「これまでもインセンティブ型の都市計画を進めてきたが、大都市以外には効力がないという問題が出てきた。インセンティブ型計画の多様化がうまくいかないと、サステナブル社会の実現は覚束ないのではないか」と締めくくった。

6: Evaluation of planning legacy and urban stock
・Overcome of negative urban stock: unplanned and unsafe urbanized area, weak area for supposed hazard
・Toward sustainable & compact urban and regional structure: land use management & renovation of existing stock, utilization of community power
Combination of leadership approach and bottom up approach

1-5　サステナブル社会のドイツのまちづくり―そのマネジメント
Sustainable Development through Machizukuri in Germany—Management and Tools

H. シュトレープ
STRAEB, Hermann

■サステナビリティは「未来の可能性を探る行為」

　本シンポジュウムのために来日したヘルマン・シュトレープ氏は、1994年から旧東ドイツのライネフェルデの団地再生まちづくりに参画し、ワールド・ハビタット賞を受賞した都市計画家である。「このような有意義なシンポジュウムに参加できることをうれしく思う」とあいさつし、サステナブル社会の基本的な考え方についてこう切り出した。「持続可能性とは、大量消費社会のなかでちょっと忘れられたのかもしれないが、昔からあるコンセプトだ。産業化の発展がすなわちサステナビリティの終焉ではない」

　ここで1枚の写真を投影。広大な土地に太陽光パネルがずらり並んでいるメガソーラーの姿が映し出された。（図1）

　「今、ドイツでも日本でも農地に太陽光パネルを設置して電力を生み出しているが、これはサステナブルなのか？」と疑問を呈した。

　ドイツにおけるサステナビリティの一般認識は、農業生産の時代からグローバル化の時代に至る今まで、その歴史的系譜によって、それぞれの時代にそれぞれの悩み事があり、時代ごとにサステナビリティの定義を変えてきた。（表1）

　「農業生産の時代は『来年も再来年ももうまく作物ができるといいね』が定義だった。しかし、世界の平和を願わなくてはならない今の時代には『こうなっていればサステナブル』という明確なものはない」と主張。つまり、将来のチャンスをもがきながら求めている姿そのものがサステナビリティなのだ。

　「サステナビリティとは一定の状況を言うのではなく、将来の可能性を探る永続的行為そのものを指す。敢然と新しいドアを開くように試行錯誤する。これが我々にとってのサステナビリティであり、我々のような都市計画者、あるいは政治家にとっての課題であり、責任である」

　実際にサステナビリティの定義は変わってきた。最初はエコロジカルに焦点が、その次にエコノミー（経済性）が付加され、さらにソーシャルライフ（社会生活）が3つめの柱として加わった。この3つの柱が環境につながっている。（図2）

■法定計画ツールで予見性を示す

　次に都市開発の話に移った。都市開発・都市計画とは、コンフリクト（対立関係）をマネージメント（管理すること）で、それを司る都市計画家は「対立意見の調整役だ」と断言する。

　対立は各所で起こるものだ。個人と個人、個人と公共、公共と公共も対立する局面はある。世界遺産を守ることと経済行為も対立関係になりがちだ。これは解決しなければいけない。

　サステナビリティとの関わりで見ると、都市は「成長」から「縮退」へとパラダイムが変わってきているため、こうした背景におけるサステナビリティについても考えなければならない。

　旧東ドイツのチューリンゲン州は、2060年までに人口が40〜50%減少すると予測されている。人数だけでなく、年齢構成も変わる。生産年齢人口（20〜65歳）が減るため、対策は必要だ。「都市計画と大きく関わる都市計画家はこうしたことまで考えなくてはならない」と語る。

　かつて、第二次大戦後のドイツのプランニングと建設の法制システムは、壊滅した都市環境の復興を優先するものだった。それによって迅速な経済の発展と豊かな生活の条件を整えた（図3）。この時代には、

1. "Sustainable development" got larger popularity in Germany, when the Club of Rome published in 1972 his report "Limits of growth". The focal point was how to provide the growing population on earth with food and goods, without destroying environment and the limited natural resources an preserving them for future generations. In this spirit, legislation in Germany started fixing critical values to limit pollution of air, soil and water in order to hinder damages of human health and to protect the environment.

2. Rather soon, it became evident, that this approach does not reflect the complexity of the interaction between man and nature. In fact, we have not only the conflict between nature and destructive exploitation of natural resources through mankind, we have competing ecosystems and we have conflicts of interests between different target groups. Sustainability cannot be achieved through definition and control of limiting values, sustainability is not static, it is a permanent struggle for balance. Actually, the current model to explain sustainability is the "Three Column Model". It integrates ecologic, social and economic development as constitutive factors. Urban development and urban planning reflect in an exemplary way the evolution of understanding of the complexity of sustainability.

3. After world war II, planning and building legislation in Germany had first of all to enable reconstruction of the destroyed cities and create conditions for a fast growing, prosperous economy. In this period, environment was rather seen as an obstacle to economic development, and social development was seen directly linked to prosperity.

　The official development tools were the land use plan covering the boundary of the municipality and the development scheme for selected zones of development within the municipality. The authority to plan future development through these instruments was given directly to the municipalities with the obligation of harmonisation with superior planning levels.

　As a very important element within the process of planning, there was introduced the obligation to present the project during 4 weeks in public and to consider eventual remarks. The idea behind this participation, to create a balance between public and private concerns and to avoid later conflicts.

4. In the meantime the understanding of sustainability has improved and contributed to first changes in legislation.
The introduction of critical values for the pollution of air, water and soil helped to reduce dangers for health and environment. But avoiding pollution is only a small contribution to sustainability.

5. By the beginning of the 90th of the last century, the discussion on sustainability developed the "three column model". Its basic idea is to define sustainability as a balance between ecologic, economic and social criteria, a balance that has to be negotiated, controlled and established in a permanent process.

6. According to this extended understanding, planning legislation has introduced the obligation to analysis for every project of major dimension the ecological impact and to define compensation measures.

　The strategy imposed by the law on protection of nature and natural resources is:

環境はむしろ経済発展の障害とも受け取られていた。法制としての都市開発手段（ツール）は、自治体の領域を対象とする土地利用計画（図4）と選定した地区を対象とする建設ガイドライン（地区詳細計画、B-Plan／図5）の2つが用意された。

パラダイムシフトが起きているなか、シュトレープ氏は新しいプランニングツールを発見した。ただし、縮退の局面でどう用いるかはテスト中だと言う。

ドイツのまちづくりシステムがサステナビリティの実現とどうかかわっているかを考えると、自治体が都市計画行政上の責任をもっていて、都市計画が地方分権され、国土のすべてを覆っていることは、先ほど大村氏が話したように日本に比べ優れた点だろう。さらに、法定計画ツールをもとに「5年後、10年後にはこうなる」と投資家や国民に対して法的予見性を与えている点、参加と利害のバランスをとって「対立を管理していること」もドイツの特徴といえる。

■都市計画に組み込まれた環境保護

近年は、計画法と検討の手続きのなかに「環境」の側面を入れることが統合化された。つまり、環境保護を計画の段階から取り組むことが求められているのだ。また、持続可能な開発に関して財政的な優遇策がある。郊外にはなんらかのかたちで助成金が導入され、持続可能な開発が可能となっている。

こうしたことができるのは、まず「地域発展計画」があり、その下に「土地利用計画」があり、さらにその下に「地区詳細計画」があるというドイツの仕組みによる。自治体が権利をもち、具体的かつ効率的な土地利用のマスタープランをつくり、地区詳細計画につながるのだ。日本とは少し違う。

都市計画を立てる際、ドイツでは環境評価をしなければいけない。植物、動物、土壌、水、大気、気候、騒音、レクリエーション価値などに悪い影響を及ぼすと判断された場合、第一の手段として「干渉を避ける」。どうしても避けられない場合は「最小化」の努力をする。それも困難な場合には、同じ敷地内での干渉を最小化するプランを探る。それでも解決できない場合は、その自治体のエリア内で対応するという「埋め合わせ」が法的に定められている。

エコシステムのバランスに影響する建設行為をする場合は、計画対象エリア内で「贖うこと」が義務づけられていることは興味深い。その実施はプロジェクト主体に義務づけられ、これらによるコストアップは、自然資源利用の経済性向上のインセンティブにもなる。

■人々が参画したマインツの路面電車

ドイツのプランニングシステムには利害関係の管理の手立てとして「参加」が導入されている。人々は都市計画を見て「好き」「嫌い」と言うことができる。しかし、イエス、ノーだけでは意見やノウハウは吸い上げられない。そこで、優れた事例として、マインツ市でのトラムシステム（路面電車）新設の際の統合的「参画」プロセス（合意形成システム）を紹介した。

マインツで延長30kmの路面電車をつくろうとしたとき、「そんなものは必要ない」という声があがった。そこで「参加のプロセス」を市民が求めた。対立を最小化して、市民に認めてもらおうという計画がはじまり、建設的な意見が交わされた。

この過程のなかで、まず住民の「ニーズ」が洗い出された。投資家もさまざまな事情をはっきり把握できるメリットがあった。特にワーキンググループをつくり、毎回同じ人が参加することにしたことはきわめて重要だった。

ワーキンググループは実際に現地（30km）を見て回った。マインツを代表するような団体も政治家も参画したので、住民がクリアなかたちで

1. Avoid interferences
2. If not possible: minimize interferences
3. Compensate interferences in situ
4. If not possible: Compensate within boundaries of the municipality.

7. The compensation measures are integrated part of the project, their implementation is obligatory. By the way: the increasing prices proved to be a very effective incentive for economical use of natural resources. If we can see some improvements in the field of environment protection, there is no relevant changes in legislation on participation.   Participation is obligatory – as ever – for the establishment of the formal planning instruments Land-Use-Plan and local development scheme. According to the phases of planning, two steps of participation are imposed:
    Phase 1: preliminary design
    Phase 2: final design.
  In both cases, the public has the opportunity to comment on a plan and to express both objections and suggestions.

8. In recent years, we have seen, that this kind of participation is not sufficient. The critical point is, that the actual system of participation only asks, if people accept a project or not. The answer comes in a very late phase of design, where changes are not welcome and expensive. We can observe that people become hostile to changes, since the project cannot be understood in its strategic context.

9. Experiences with a more extensive understanding of participation prove, that early and intensive participation can contribute to a higher quality by the mobilization and use of the local and technical knowhow of the stakeholders and it helps to improve the chances for implementation because of better acceptance. Based on these positive experiences, the legal framework for participation in Germany is actually on trial.

10. As an outstanding example and prototype of this new approach the integrated planning and participation process for a new tramway in Mainz, Germany will be presented.
  But there is not only the need to improve participation. We also have to have a critical look at our legal planning instruments. Both Land Use Plan and Local Development Scheme are static instruments, as they define with rather high binding force a final situation. This was acceptable in times, where demographic and economic growth offered interesting investment opportunities.

  But in times with shrinking population and economic stagnation, the investor becomes more demanding and wants to have an impact on the content of the formal planning tools. This means, that development is a process, where you don't know in the beginning, what will be happening. In this case, it is indispensable to clarify the objectives in order to have clear criteria for project decisions. For this reason, we have to establish strategic, integrated development plans helping to manage the development process rather than fixing a final picture of development. In eastern Germany, which is particularly affected by shrinking problems, these Integrated Strategic Urban Development Plans have been introduced in 2002 as base for allocation of subsidies. An example combining the Integrated Strategic Development Plan with a new dimension of participation will be presented by Mr. Miller, Gera.

11. In order to promote the idea of sustainable urban development, a national prize for sustainability is awarded every year to the municipalities with best practices. The criteria for this price reflect the complex understanding of sustainability
- Governance & Administration,
- Climate & resources,
- Mobility and infrastructure
- Economy and labour
- Education and integration
- Quality of live andurban structure

政治にアクセスすることもできた。参加者がどこに住んでいて、何を心配しているのか把握することも努めた。政治的なレベルの話はしないことを決めてディスカッションに臨んだ。

30人が参加した最初のワークショップでは「何を恐れているのか?」と潜んでいる問題点を抽出。そこを出発点として、プランナーがプランを見直し、目標の実現を評価するなかでソリューションをつくった結果、経済性がよく、投資も抑えた、住民が喜ぶトラムシステムとなった。「都市計画家としては建設許可が下りやすかったし、すべての住民が知るプロジェクトとして続いたので、投資家も満足した」と述べた。

住民が参画することによって、当初のプランがだいぶ変更された。図6を見てほしい。黄色がもともとの計画で、赤と緑が最終的に実現したプランだ。かなり変わっている。「当初、トラム会社はもっと先まで延ばしたかったけれど、住民の話から実態がわかってきて『ここは誰も使わないから投資しなくてもいい』と400m短くした」と振り返る。(図6)

初期段階でより深く参画してもらった方が、利害関係者のもつ地域の技術や知識、ノウハウが発掘できるうえ、プラン(プロジェクト計画)のクオリティ・アップにつながる。また、地域から受入れられやすくなり、計画の実施も有利になると言う。

■包括的なプランニングと心理的なインセンティブ

「まちづくりシステムをサステナビリティの実現と結びつけるには、持続可能な事業のための助成金をつくるべき」。これがシュトレーブ氏の考えだ。統合的な開発コンセプトに対する助成金はドイツにもあるが、人口が減っていくなかでは意味がない。従来とは違うプロセスから、社会的背景と環境も含めた都市開発に変え、もっと包括的なプランニングにすることが求められている。ドイツでは都市再生、気候変動や再生可能エネルギーにも補助金を出しているそうだ。

一方、心理的なインセンティブも必要だ。ドイツには持続可能な「ドイツ・サステナビリティ賞」という制度が始まっている。持続可能なことを国がどれくらいのレベルでチェックしているのかわかるおもしろい制度である。地域自治体が応募する制度で、ベストプラクティスを競うもの。6年前に発足し、3年前からは都市開発コンセプトも対象になっている。審査基準は、サステナビリティの複合性が反映されて設定されている授賞のカテゴリーには、ガバナンス、気候変動対応、資源問題、モビリティ、経済活動、教育、クオリティ・オブ・ライフなどがある。このカテゴリーだけを見ても、ドイツ政府がサステナビリティということに、いかに包括的なアプローチをしているかがわかるだろう。

シュトレーブ氏は、最後に「サステナブル・アーバン・プランニング」のガイドラインを紹介した(表2)。まずは、まちをプロセスとして考える、前向きに考えることが大事だ。そして、なにが問題なのか、その実態をはっきりさせること。ステークホルダーをまきこみ、戦略的に考えながら目に見えるメッセージを発していくこと。それぞれの状況が違うことを理解して、画一的な計画は避けることなどを主張。そのうえで「サステナブルは環境の再生であり、建築物のデザインクオリティも忘れてはならない」と述べた。

図1 What is sustainable? これがサステナブルなのか?
Fig. 1 Sustainability achievement requires more than solar cell

表1 ドイツにおけるサステナビリティの一般認識
Tab.1 Development of sustainability ideas in Germany

歴史的系譜	主な関連事項	Historical context	Major concerns
農業生産の時代	地域消費の農産物の生産、商業活動	Rural period	Food production for local consumption and trade
工業生産の時代	資源の入手、市場の形成、ハウジングの形成そして人口の増加	Industrialisation	Availability of raw materials and markets, housing for fast growing population
第二次大戦後の時代	復興建設、再構築、経済活動の再生、社会福祉	Post world war II	Reconstruction, economic re-start, wealth
ローマクラブの報告の時代	成長の限界 環境汚染の管理 自然資源の保全	Club of Rome	Limits of growth: Pollution control Preservation of natural ressources,
グローバル化の時代	気候変動 再生可能エネルギー 社会と世界の平和	Globalisation	Climate change, renewable energy, social and global peace

図2 サステナビリティを支える3本の柱
Fig. 2 Three pillars support the achievement of sustainability

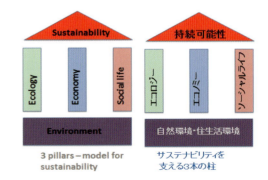

図3 地域発展計画図
Fig. 3 Regional development plan

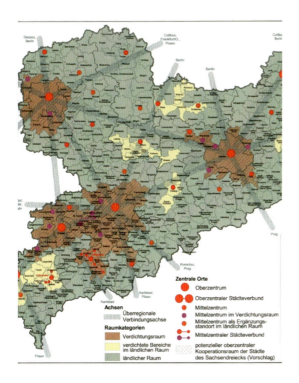

図4 土地利用計画図
Fig. 4 Land Use Plan

図5 地区詳細計画（Bプラン）
Fig. 5 Building regulation plan / Bebauungplan

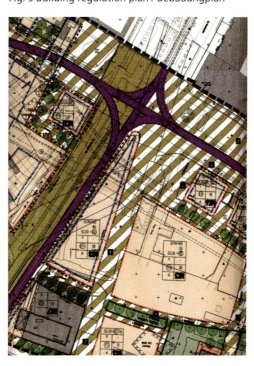

図6 マインツ路面電車計画―住民と検討した路線計画案
Fig.6 Some ideas in the planning process of the transit system

表2 マインツ路面電車計画―合意形成ガイドライン
Tab.2 Discussion guidelines for sustainable urban planning

Think the city as a process まち"を"まちづくり（プロセス）として考える	**Facilitate implementation** 実施体制を整える
Think positive ポジティブ（前向き）に考える	**Transmit visual and touchable messages** 見える、触れるメッセージを伝える
Identify the real problem 問題の実態をはっきりさせる	**Insist on quality** クオリティにこだわる
Integrate the stakeholders 利害関係者を統合する	**Cooperate and integrate** 連携と統合化を図る
Think strategic 戦略的に考える	**Recognize the singularity of the situation** 状況ごと特有性があると考える

1-6 討議とまとめ:
プロジェクト・アプローチ、プランナー・コーディネーターの役割
Discussion and Conclusion – Project approaches, and roles of planners and coordinators

澤田 /SAWADA、大村 /OMURA、シュトレープ /STRAEB、猪俣 / INOMATA、中上 /NAKAGAMI、永松 /NAGAMATSU

■「サステナブルなまちづくり」とは

大村 内田発言は適切だ。「使い続けられる建築」は、ニーズの多様化への対応可能ということであり、現在のまちづくりが建築、地区、都市の全レベルで変わらなければならなくなっていることに関係づけられた。

シュトレープ 内田発言の示唆したのは、「成長の時代」から「成長しない時代」のパラダイムシフトなのだ。つまり"何をつくるかを考える"のではなく"何の変化なのかを考えるべき時代"に入っている。したがって、その変化に対応するにはどんな可能性を持たせれば良いかを考える必要がある。今の時代は「解体して建設」は受入れられない。プロジェクトのプロセスでは投資計画が決め手になる。その際には建物がどうしても必要かどうかを予想し、30〜40年経った時に違う用途で使えるようにするのかどうかも、考えることになる。つまり、"つくる建物"に「順応性」が必要であり、都市にレベルを上げても同じことなのだ。

大村 井上発言の中で、「現状は一気には変えられない、ストックを大事にしつつ、同時に不要になったものを順に畳んで行く」という話しがあった。しかし、実際のまちづくり活動を見てみるとこの方法は簡単ではない。サステナブル社会への移行は簡単ではない。

■サステナブル社会のまちづくりシステム

澤田 まちづくりという活動が日独では違う。ドイツでは順に意思決定を進め、プランをつくる。ところが、日本ではまず計画案をつくり、利害関係者がそれについて議論する。サステナブル社会のまちづくりでは、「長く使い続けられる建築」を考えることはたしかに意味がある。

ならば、この「可変性・柔軟性のある建築や街」を具体的にどう計画し実現するのか?つまり既存の建物や街の環境とどう馴染ませるか?成長しない時代という変革期にあって、まちづくりの専門家はこの社会変革をどう認識し、まちづくり活動をどう進めればよいのか?

猪俣 「成長の限界」が言われた時代にすでに「使い続けること」の重要性が認識され、様々な分野での「使い続けること」に関する検討がはじまった。まちづくり分野でのこの問題の検討はようやく今になって始まった。ずいぶん遅れていると思う。

永松 そもそも日本人にはサステナブル社会という認識があるのかどうか、自分は疑問に思うことがある。自分は、東北で暮らしてまちづくりの委員会に参加しているが、そんなことを考えさせられる。

中上 エネルギー問題に取組んでいる自分の立場では、従来のプランニングが通用しなくなった状況があることは十分に認識している。エネルギーシステムはサステナブル社会づくりの核になることだが、そのプランニングは、日本では大変難しく、絶望的な気がしている。

澤田 エネルギー問題も社会変革の最重要課題だ。一般にドイツでは、エネルギーの生産と貯蔵、供給が大事で、それで完成するという認識だ。しかし自分はもっと分散化と分権化が必要だと思うし、日本から学ぶことも多い。日本は暖房用エネルギーの消費が少ないが、ドイツは沢

■ The Concept of "Sustainable Machidukuri"

OMURA Professor Uchida makes some good remarks, the constant use is a response to the growing variety of needs, and it has become an age where change is required at all levels of the built environment.

STRAEB It is a paradigm shift from a "growth period" to a "non-growth period". We must think of what we should change rather than thinking of what we should build.

Nowadays the "scrap & build" mentality cannot be accepted and therefore we me must plan well the construction process in order to invest. We must ask ourselves if the building is truly needed and if it can be used 30-40 years after being built. What buildings need is "flexibility" and the same can be said for our cities.

OMURA As Prof Inoue stated "the current situation cannot be changed drastically, the process of protecting the stock while deconstructing what is unnecessary" has proven to be a difficult task.

■ The Development of a Sustainable Society

SAWADA The Japanese concept of machidukuri is different from the one in Germany. The German concept follows a decision making level process, whereas in Japan we follow a plan proposal. This means for a sustainable machidukuri we should reconsider "architecture which can be used in long terms". We must consider how to accomplish flexible architecture/cities and in order to do so we must ask ourselves how to adapt to the existing environment. The current society is going through a transition and I would like to ask how other experts are affronting this transition.

IINOMATA Our era is an era where we have met our "growth limits" and an era where we have started to acknowledge the importance of the "constant use" in many fields, even in the machidukuri field the concept has finally started to be considered.

NAGAMATSU I am a North East Japan member of the committee and my experience there has led me to think that perhaps we must first ask ourselves if the Japanese society has any perception of livening in the age of sustainable societies.

NAKAGAMI From an energy issue perspective, our current methods have become inefficient and I find it difficult for it to become the core for sustainable society development.

SAWADA he energy issue is a problem which has to be addressed as part of the machidukuri and is a major point in the transition of a society.

The production, storage and supply of energy are an important matter in Germany and therefore it is a matter which is incorporated within a scheme generally. Although I believe that a decentralisation is needed, there are many aspects which can be learnt from Japan.

By the way the NRW prefecture Bottrop project that will be introduced this afternoon will be a project that will focus on the relationship of energy transition and machidukuri. It is an example of a machidukuri project which has a clear project process and criteria. The heavy industry has historically supported the social development

山断熱材を使い、同時に湿気も溜めてしまっている。

ところで、午後の部で紹介されるNRW州・ボトロップのプロジェクトはエネルギー変革とまちづくりに関することだ。プロジェクトのプロセスとクライテリアが明確に設定されているまちづくりの事例だ。

あの州は重厚長大産業で社会発展を支えて来たが、IBAエムシャーパークでよみがえった地域なので、社会変革とまちづくりとを合わせて考えて来た経験がある。

■まちづくりプランナーの役割り

澤田 もう一度ドイツと日本でのまちづくりプランニングの比較に戻りたい。ドイツのまちづくりでプランナーの仕事は割が合うのか？報酬はどのように決まり、今後の具体的な業務はどうなるのか？東西統合からもう25年経ったが、その社会変革の過程でこの点はどうなのか？

シュトレープ 仕事は変化してきている。いろんな要素をいかにインテグレートすべきかに変わって来ている。したがって常に学び続けなければならない。仕事はある計画（プロジェクト）に関して引受け、報酬額も決まる。日程や対象エリア規模などは固定しない包括的契約だ。これが建築設計の業務とは違う。しかし、建築家と違いプランナーは10年さらに20年とクライアントに対応しなければならない。プランナーに課せられた使命があるので、がんばるしかない。

また、自治体がクライアントであれば、必要なデータは統計局を通じて手元に入る。調査対象を特別な場合に自分たちがインタビューしなければならないこともある。また個人情報の入手は難しい。

東西統合の後は、土地所有など実態の変更が増えて来ていて、政府のデータで裏付ける必要が発生している。自分たちのチューリンゲン州の政府にもモニターの部署が出来ている。

都市計画の業務は複雑な内容である。論争一つで完結するわけではない。関連事項が沢山ある。プランニングの効果があるかどうかを筋道立てて説明できるようにするのは難しいものだ。

大村 日本におけるプランニングでは、計画案ができた時、住民に2週間縦覧することが法的に決められている。しかし、それでは形式的過ぎるので、住民の身近な環境・施設のプランニングで地権者などの関係者の参加も増えている。

永松 自分は東北のまちづくりに関わっているが、こうした合意形成の進め方は3.11の前とその後ではずいぶん違って来ている。東京など大都市では合意形成プロセスを慎重に進めているのに対して、東北では、住民利益を考えないわけではないが、このプロセスに掛かる費用と時間の方を重要視しているようだ。また、住民もそれを悪いとは思っていないと自分は見ている。

東北のそれほど被害の無かった地域は、人口減少との関わりでけっこうナーバスになっているように見える。

だから、こうしたシンポジュウムに参加して、仕事ぶりを振り返る意味がある。

of the NRW prefecture and therefore there has traditionally been a close relationship between the society and its machidukuri.

■ The Job of a Machidukuri Planner

SAWADA I would like to go back to the comparison of the machidukuri planning of Japan and Germany. Does the job of the machidukuri planner have a good salary? How are the fees worked out and how will the role change in the future? It has been 25 years since the unification of Germany, how has the change in the Germany affected the role?

STRAEB Jobs change in order to integrate many factors and therefore we must continue to study and learn. Once we get an offer for a project we decide on the economic terms. In contrast to architectural design work this is normally done without taking into account the schedule or the target area. One of the obligations of the Planner is the obligation to have to work in close proximity with the client for 10-20 years. Therefore we must make effort to keep a good relationship. When the local government is your client data can be gathered by statistic offices. Once the target can be identified we do personal interviews with these targets, at times the handling of personal data can be difficult.

After the unification of Germany, the government was required to verify its data due to the increased changes in data. Even in Thuringia we have a government monitoring agency which gathers data. The subject of urban development is a complex matter which revolves around many factors. Therefore it is difficult to logically justify the planning effects.

OMURA When planning in Japan the public have a 2 week period where they can examine the plan proposal by law. But many think this method is too formal and try to include the public by inviting them to be more involved during the planning process.

NAGAMATSU I am involved in the machidukuri of the Tohoku and think that the perception of the governing body has changed since the incidence of the 3.11. In large cities such as in Tokyo the governing process is held with care. Although regions such as the north east tend to focus more on the time and resources spent and the public have no objections regarding this perspective. This is mostly due to the pressure which is exerted on the governing body in trying to recover from the large earthquake. Even regions of Tohoku which were not heavily affected by the incidents have suffered a population decrease. This population decrease has made the people nerves regarding the matter of reconstruction.

I think this symposium is a good opportunity to observe and compare how the term of sustainability is treated differently in densely populated Japanese regions, low density Japanese regions and compare them to German regions.

第2章　サステナブル社会のまちづくり―日本とドイツにおけるプロジェクト
Chapter2　Sustainable Machidukuri Projects in Japan and Germany

ノルトラインヴェストファーレン（NRW）州とチューリンゲン州
North-Rhine Westphalia (NRW) and Thüringen States

会議の開催地、プロジェクト事例の所在地
Conference Venues and the Projects to be presented

東京会議 2
Tokyo Conference 2

G. ロエル（文責：前川）
Mr. ROEER
イノベーションシティ・ボトロップのまちづくり
Innovation City - Bottrop Machidukuri Project

H. シュトレープ（文責：前川）
Mr. STRAEB
IBAチューリンゲン 2013-2023
地域再生プロジェクト
IBA Thüringen 2013 – 2023
Regional Redevelopment Project

奥茂賢仁（文責：前川）
Mr. OKUMO
日本の団地再生・まちづくりの現状と課題
Recent Housing Refurbishment Projects in Japan

松村秀一（文責：前川）
Prof. MATSUMURA
「箱」の産業から「場」の産業へ―
まちづくり産業の転換
From "Box"-making to "Place"-making industry –
Change of Machidukuri Industry In Japan

R. ミラー（文責：前川）
Mr. MILLER
ゲーラ2030まちづくりプロジェクトの仕組み
Gera 2030 Project –
Integrated urban development project

中村圭勇（文責：前川）
Mr. NAKAMURA
富山市-環境未来都市-コンパクトシティ・コンセプト
Toyama City Environment Future City & Compact City

小玉祐一郎（文責：前川）
Prof. KODAMA
パッシブタウン黒部モデル
Passive Town Kurobe Model and Machidukuri

討議とまとめ - 日本の成長産業、
コンパクトシティのプランニング、
合意形成の実態
Discussion and Conclusion –
Urbanization, compact city concept and consensus building

2-1　イノベーションシティ・ボトロップのまちづくり
Innovation City - Bottrop Machidukuri Project

G. ロエル
LOER, Georg

■「環境未来都市」への転換めざす

　株式会社NRWジャパン（ドイツ ノルトライン・ヴェストファーレン州経済振興公社日本法人）の代表取締役を務めるゲオルグ・ロエル氏。再生可能エネルギーやロボティクスを重点に、日独プロジェクトのコンサルタントなどを務めている。ロエル氏は流暢な日本語で、イノベーションシティ・ボトロップ市（以下、ボトロップ）のまちづくりについて講演した。

　ボトロップはドイツのノルトライン・ヴェストファーレン（NRW）州にある人口約12万人の都市だ。ルール工業地帯の中心地の1つである。（図1）

　「ボトロップにはまだ炭鉱が残っている」とロエル氏が語るように、ボトロップの経済は何十年にもわたって石炭鉱業によって支えられてきた。現在ドイツで稼働中の最大規模の石炭鉱山も同市にある。しかし、ドイツの石炭鉱山の段階的廃止政策（ボトロップ石炭鉱山は2018年閉山予定）によって、ルール地方全体が厳しい構造変革にさらされている。

　こうした背景から、ボトロップでは「都市をいかに活性化するか」ということが1つの大きなモチベーションとなっている。また、日本でもゲリラ豪雨が頻発するなど異常気象に見舞われているが、ボトロップにも気候の変化が見られ、洪水が起きることもしばしばだ。（図2）

　「こうした時代背景のなかで、ボトロップというまちとして、さらにはルール工業地帯という地域として、どういう風に対処できるかが大きな課題となっている」と述べる。EUは地球温暖化に対する目標を掲げている。また、ドイツは2020年までに原子力発電所をすべて止めると決定した。

　ボトロップは、ドイツ連邦政府が原子力発電所の廃炉を決定する以前の2010年から、ある取組みをスタートしている。それがここで紹介する「Innovation City Ruhr（イノベーションシティ・ルール）」だ。

　イノベーションシティ・ルールとは、2020年までに「環境未来都市」に転換するまちを選ぶコンペティションで、第三者機関による最終審査からボトロップが選ばれ、2010年にスタートしたもの。ボトロップの居住地域を、省エネルギーと持続可能性に適合するまちに変換すること、それによって2020年までにCO_2の排出量を半減させて、クオリティ・オブ・ライフ（生活の質）の向上をめざす画期的なプロジェクトである。

■レトロフィッティング（改修）を前面に

　イノベーションシティ・ルールの重要なファクターの1つは「市民によるサポート」だ。東日本大震災の影響もあり、2011年以降、急速に進むエネルギーシフトを遂行できるかという課題に、ボトロップは市民のサポートを得ながら取り組んでいる。すでに、市内の人口12万人のうち7万人、建物数1万4500棟（うち住宅1万2500棟、商業ビル2000棟）を擁する居住区を「環境未来都市」にふさわしいまちとするため、複数のプロジェクトをスタートしている。（図3）

　イノベーションシティ・ルールの構造変革の総合コンセプトは、ボトムアップでプランニングして、個々の建物レベルから地区レベルへと進め、最終的に市全体に広げるというもの。人口7万人の居住区をモデル都市として、優れた点、困難な点などを学び、ボトロップという1つのまちの施策をルール地方のほかのまちに広げていくことを考えている。

　「さらにノルトライン・ヴェストファーレン州全体に、最終的にはドイツ全土での取り組みに育てたいという目標を抱いている」と述べた。

　また、エネルギーそのものに関しても「ボトムアップ方式」を重視する。

■ Innovation City Ruhr and Bottrop project

InnovationCity Ruhr is a landmark project aimed at realizing the conversion of an existing residential town quarter of the city of Bottrop into a housing area geared at energy conservation and sustainability thus cutting CO_2 emissions by half until 2020 and raising living quality in the area.

Bottrop is located in the German state of North Rhine-Westphalia with almost 120,000 inhabitants. Being part of the so called "Ruhrgebiet", for decades Bottrop's economy has been shaped by coal mining with today's biggest and still active German coal mine also being located in Bottrop. However due to the phasing out of coal mining in Germany and Bottrop's coal mine set to shut down in 2018 the whole area has been undergoing profound structural change for decades now.

InnovationCity Ruhr set out in 2010 with a competition to select a "typical" Ruhrgebiet-residential area in order to turn it into the "climate city of the future" by 2020. Among the several finalist Bottrop was finally selected by an independent jury.

Several projects have since been started to develop a residential quarter of Bottrop with about 70,000 inhabitants into a blueprint for a climate friendly city of the future. The general idea of the conversion project is a bottom-up approach, progressing from the level of individual buildings to district level and finally encompassing the whole city. The goals of InnovationCity Ruhr are in line with the general project of the German Federal Government of shifting German energy production to renewable energy sources and phasing out nuclear and coal fired plants, a process termed the "Energiewende" (energy transition).

Until today over 200 different conversion projects have been implemented in the areas of "Living", "Working", "Energy", "Mobility" and "City (planning)". Each of them is geared towards targets like energy conservation, shifting energy production to renewable sources or creating an infrastructure for smart energy production. It is important to stress though that this conversion must not be misinterpreted as a sort of de-industrialization. Quite contrary local businesses actively take part in the initiative and strive to also realize the economic benefits of saving energy. Additionally new jobs have been created in innovative and "green" technologies.

InnovationCity Ruhr's approach and successes have drawn attention from various outsiders. Recently InnovationCity Ruhr has been granted a special award by the National German Sustainability Award competition.

■ National German Sustainability Award

Since its inauguration in 2008 the National German Sustainability Award has developed into one of the most prestigious European awards of its kind. It is aimed at honoring outstanding achievements in the promotion of the idea of sustainability, which can be defined as the acceptance of social and ecological responsibility across society. The awarding ceremony is a large media and social event, thus drawing further attention to the individual projects and raising overall awareness for sustainability. The award is being endorsed by the German Federal Government, local and business associations as well as numerous NGOs, among them UNESCO and UNICEF. In several categories sustainable projects are submitted and – after undergoing a shortlisting process carried out by outside consultancies – are finally reviewed by a jury consisting of 15 independent experts from various fields. The multi-level selection

熱、エネルギー、ガスの3つをうまく利用して、エネルギー効率をよりよくすることが必要だが、「その狙いを達成するためにはいろいろな最新のシステムを使うことを考えた」と話す。地域暖房や最新の風力発電などをさまざまな建物に展開するほか、エネルギーマネジメントにも挑んでいる。エネルギーをストレージするシステムをつくり、例えばバスに用いるといった施策も講じている。（図4）

ボトロップは、現在年間24万6000トン排出しているCO_2を、2020年までに50％削減するという数値目標を掲げた。中国などの新興国とは違って、既存のまちをいかに効率よくエネルギーを使うために変えていけばよいのか。これに対して「レトロフィッティングが有効だ」と言う。この場合のレトロフィッティングとは、都市の改造や市街地の改修、建物のリノベーションなどを指す。古い建物を改修し、エネルギー効率をアップして、CO_2排出量を削減しつつある。まだ中間報告だが、ボトロップの2012〜2013年の改修率は7.82％にのぼる。ドイツ全体の改修率が2012年で0.9％であることを考えると、相当高い数値だとわかる。（図5）

■こまめな情報提供で理解を得る

ルール地方におけるイニシアティブから生まれたイノベーションシティ・ルールは、いくつかの都市が競合するなかボトロップがもっとも優れた提案と認められた。今、ボトロップでは、次々とプロジェクトが実施されている。

「これについては、先に述べたように市民を取り込むことは大事で、さらに企業の方々を取り込むこともきわめて重要だ」と主張する。（図6）企業のほかにもアーヘン工科大学などルール地方内外からアドバイスをもらい、ノルトライン・ヴェストファーレン州でもさまざまな省庁がサポートしている。いまや住居、仕事、エネルギー、モビリティー、都市全体など、全部で200におよぶプロジェクトが動いている。

ここでロエル氏はいくつかのプロジェクトを挙げた。

まずは住居から。いろいろなかたちの住居や商業施設があるなかで、1つの模範的な例を示し、分譲住宅、マンション、商業施設それぞれで目標を達成するためのエネルギー効率向上の施策をとっている。

低所得層が住む住宅公社は「集合住宅プラスエネルギーの建物」を建てた。これはエネルギーを消費するだけでなく、逆に供給することをめざすもの。

大学でもプロジェクトが動いている。「エネルギーキャンパスラボ」として、エネルギー効率の向上に加え、エネルギーをつくり、溜めるといったいわば「ショーケース」と位置づけたもので、2014年9月に完成済みだ。

ノルトライン・ヴェストファーレン州が奨励している固定型燃料電池のプロジェクトも、ボトロップで一部実施することになった。GWIというガスの専門的な研究会社が、ボトロップで燃料電池、あるいは最新型の暖房施設を導入することを手伝っている。

「午前中のシュトレープ氏の話にもあったように、住民もしくはステークホルダーにこまめに情報を提供することはとても大事だ」と言う。

例えば、夕方にイベントを行なって暖房、断熱、太陽光発電、資金調達を紹介し、あるいは講習会を催すなどして、住民たちとコミュニケーションをとりながらコンサルテーションやアドバイスにつなげているのだ。そうした活動が、先述の改修済み建築物978（居住用建築物全体の7.82％）という成果に表れているといえる。

■たんなる脱産業化ではない壮大な試み

このようにイノベーションシティ・ルールのプロジェクトは多方面で動いている。計画から設計、イノベーション段階へと着実に進んでいるようだが、「最終的にはイノベーションシティ・ルールのマニュアル（都市

図1　ボトロップ市の位置
Fig.1　Location of the City of Bottrop

図2　ボトロップ市内の洪水
Fig.2　A local flood scene in Bottrop

図3　パイロットプロジェクトエリア（エネルギー＋改修）
Fig.3　CO_2 emission reduction: 246,000 ton / year
　　　Population: 70,000 / Buildings: 14,500
　　　(Residential: 12,500, Industrial & Commercial: 2,000)
　　　Retrofit/renovation, 2012-2013: 7.82% of all buildings

図4　ボトムアップ方式のエネルギー転換
Fig.4　A scheme of energy reform in Bottrop: "bottom-up"

改修への実施設計図）をつくることが目標だ」と明かす。

そのためのステップとして、まずは現状調査がある。そして、どんなエネルギーをどこで使うか、それに適したどういう施設があるのか、さらに収益源やソーラーパワーの使い方、それぞれでどう施策を展開するかといったことを、"パートナー"の協力を得て適確にコーディネートするかたちで各プロジェクトを実施している。（図6、表1）

「イノベーションシティ・ルールのマネジメントとは、各プロジェクトを束ねていくこと。それと同時にステークホルダーを取り込むこと。これが大事だと考えている」と語った。

プロジェクトはそれぞれ、省エネや再生可能エネルギーへのシフト、スマートエネルギー利用インフラの構築などを目標としているが、この構造転換が「たんなる脱産業化」と思ってはいけない。地元企業はこの新たな取り組みに積極的に参加し、省エネの経済的利益の享受をめざしているし、革新的なグリーン・テクノロジー分野ではすでに多くの雇用も生み出している。

ロエル氏は、重要なポイントとしていくつもの歯車をうまく調整できるかどうかにかかっていると言う。

「連邦政府やEUが大量の資金を投入して、まるで実験のショーケースのようにするのではなく、ボトムアップ方式でいろいろな人や企業、機関を巻き込んで多様なプロジェクトを実施することが大事だ。企業も数多く参加し、大きな貢献を果たしてくれている」

イノベーションシティ・ルールのこうしたアプローチと成功は、各方面から強い関心を集めている。2013年には、National German Sustainability Award コンテストで「特別賞」を受賞している。（図7）

process is challenging for participants, but also enables them to critically evaluate their sustainability strategy and draw benefits from this evaluation process, even if in the end they are not shortlisted. There is also an "honorary award" handed out to celebrities who actively put their public persona to good use. Former honorary laureates include British actor Colin Firth and his wife Livia, Sendai mayor Emiko Okuyama and HRH Prince Charles.(Fig7)

表1　イノベーションシティ・ルールのパートナー
Tab.1　Project partner groups for Innovation Ruhr project

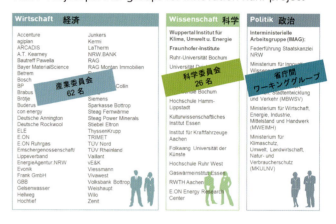

図6　一元的な地域管理体制
Fig.6　Activities of the project and management tools
　　a.consultaion on household level　b.networking
　　c.workshops　d.visualization through a mapping

図5　ボトロップ市の改修率
Fig.5　A map of buildings to retrofit/renovate in Bottrop
2012-2013: 7.82% (all over Germany: 0.9%, 2012)

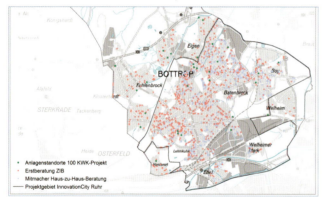

図7　ドイツ・サステナビリティ賞のトロフィー
Fig.7　The trophy of the German Sustainability Award

2-2 IBAチューリンゲン 2013-2023 地域再生プロジェクト
IBA Thüringen 2013 – 2023 Regional Redevelopment Project

H. シュトレープ
STRAEB, Hermann

■ IBAシステムは「イノベーション促進策」

ヘルマン・シュトレープ氏が再び登壇し、ドイツにおける「サステナブル社会のまちづくり」の促進策の役割を担うIBAシステムおよび現在進行中のIBAチューリンゲンプロジェクトについて紹介した。

ドイツには建築や都市のプランニングと建設の実験を一貫して進め、総合的な検証をするという伝統的方式がある。それが IBA（Internationale Bauausstellung/International Building Exhibition）国際建築展と呼ばれるものだ。1900年ごろからスタートした IBAは計画面と建築法制面の壁を取り払うイノベーションの促進策で、テーマは時代状況に応じて設定してきた。

これからのサステナブル社会の実現に向けてはイノベーションが不可欠だ。イノベーションを生むためにはインセンティブや助成金、心理的な仕掛けなど多様なものが必要となる。その観点から考えると、新しいソリューションが必要な問題を特定したうえで国際的な専門家やプロからアイデアを募り、そのなかからコンペティションで実行可能なプランを選りすぐり、かつプロジェクトの検証まで行なう IBAは、イノベーションを促進するためにきわめて有効な方法といえるだろう。しかもIBAが終わったあとも、それを実現したノウハウを一般の人たちも得ることができる。（表 1）

■ これまでのIBAの歴史とテーマ

シュトレープ氏は次にこれまでの IBAが成し遂げた代表的な例を紹介した。まず第 1回は 1901年にダルムシュタットで行なわれた「ダルムシュタット・マチルデンヘーエ」だ。これは、住宅と生活と仕事のあいだに新しい調和をもたらそうというもの。

2回目は 1927年にシュトットガルトで行なわれた「IBAシュトットガルト・ヴァイセンホーフジードルング」。バウハウスなどモダンな住宅建築のコンセプトについてのもの。当時、ドイツ国内では賛否両論を巻き起こした有名な事例だ。今も人が住み続けているので、見に行くこともできる。（図 1）

第二次世界大戦後は「復興」がテーマとなった。ベルリンの東西で破壊された中心市街地を復興する目的で「ベルリン インターバウ、スターリンアレー」が1952年と1957年に行なわれた。ここで IBAは初めて都市の再生を扱った。残存する都市インフラを活かしながら再開発をどうすべきかという事業だった。（図 2）

1987年の「IBAベルリン」のトピックは都市の再構築だ。都市のスペースを維持して、アクセスしやすくした。

そして、1999年に行なわれたのが「IBAエムシャーパーク」。鉱山が閉鎖になってしまったルール工業地帯をどう変えていくのかという「都市の再定義・再構築」がトピックだった。人口動態が変わるなか、地域全体を対象に、いろんな都市でいろんなソリューションを比較して、持続可能な将来につなげていく課題に立ち向かったものだ。昔の工場地帯をミュージアムやデザインセンターにした。さらにデザインだけでは十分ではないと判断し、工場をハウジングや科学研究機関、大学に変えるなどさまざまな再利用が提案された。（図 3）

石炭の露天掘り跡地を新しいランドスケープにしたのは2000年から2010年に行なわれた「IBAフュルストピュクラーランド」だ。露天掘りのために地表に穴が開いていたところに炭鉱が閉山となり、地下水が

■ IBA System: planning/realization/verification

Experimental town planning and architecture have a strong tradition in Germany. Since 1900, they have existed as punctual, thematically focused events alongside the rather strict planning and building regulations that are often hostile to innovation.

The tool is the IBA, "International Building Exposition" in English. (Tab.1)

Its basic idea is to focus on present and future problems and to mobilize the top international creative personalities to develop innovative solutions that are of very high quality. The process draws the attention of exceptionally qualified professionals, and the built results are presented to the public as a real-time exposition. Of course, the results are intended to be used not only for testing purposes and exposition but to prove their long-term fitness for the intended function, broad effect, and sustainability, central concerns of every IBA.

According to the historic challenges in urban and architectural development, the previous IBA had very different themes, from housing solutions to urban renewal strategies to regional transformation processes. A constitutive element of the IBA is the search for results of the highest quality through competition and consistent quality management. Organization of the IBA and quality management are committed to a limited liability company, whose shareholders are the involved public structures (ministries, municipalities).

Previous IBA sites include
 - Mathildenhöhe Darmstadt 1901
 - Weißenhofsiedlung Stuttgart 1927 (Fig.1)
 - Interbau and Stalinallee 1952 / 1957 (Fig.2)
 - IBA Berlin 1987
 - IBA Emscher-Park 1999 (Fig.3)
 - IBA Stadtumbau 2010
 - IBA Fürst Pückler Land 2000 – 2010 (Fig.4)
 - BA Hamburg 2013 (Fig.5)

■ IBA Thüringen 2013 – 2023 Project

The IBA Thüringen 2013–2023 deals with a very specific problem of Thüringen but I think it provide answers to current development issues in other parts of the world.

The settlement pattern of Thüringen is characterized by a large number of small cities located in a rather rural setting. As in all other parts of Germany, the population is shrinking. It declined from 2,611 mil. inhabitants in 1990 to currently 2,161 mil. inhabitants. The quite realistic projection for 2030 is 1,842 mil. people. The long-term projection for 2060 is 1,324 mil., just half the population in 1990.

The specific problem is that since 2003, rural areas have been more affected than small towns. The phase of suburbanization just after reunification is now being followed by depopulation of rural areas.

The situation has become more serious and is now endangering the survival of the villages, since services and infrastructures cannot be made available below a certain number of customers. Urban and rural areas have to develop new relationships. (Tab.2)

IBA Thüringen deals precisely with that relationship between urban and rural structures. The goals are to
 - link the cities with the rural areas
 - cultivate the energy transition
 - control the consequences of demographic change
 - strengthen the Thuringia identity

戻ってきた。こうしてできた湖のランドスケープを活かし、湖上ハウスや湖畔の別荘、あるいはイベントセンターなどを建設した。いずれもコンペティションによって優れたプランとソリューションを実施した。（図4）

2013年の「IBAハンブルグ」は、川の両側に分かれていたまちを、橋をかけることでつないで都市化を促した。また、外国からの移民たちをどう受け入れるべきかというトピックもあった。今、ドイツでは移民を市民として社会にどう融合するかが課題になっているが、職・住統合によって地域経済に参画させ、有機的に統合する道を探している。（図5）

■進行中の「IBAチューリンゲン」

ここからは、2013年から2023年にかけて実施されている「IBA チューリンゲン」について「今はまだ『いい方向性を見つけよう』という提起の段階にあるので始まったばかりだ」と述べた。

チューリンゲン州は、ドイツの他の地域と同じく人口が減っている。東西統合時の1990年には281万人だったが、2030年に184万人、2060年には1990年の半数以下の132万人になると予測されている。2003年以降、ルーラルエリア（田園地帯）では東西統合が影響して、中心市街地よりも人口減少が著しい。適切な生活サービスやインフラ整備といった公共のサービスが受けられないような状況すら生まれている。郊外をどう生き残らせていくか、そしてこれまで提供していなかったサービスを与えつつ中心市街地がどう生き残っていくか。「ルーラルと都市の新たな関係」が求められていることがトピックだ。

「今の世界各地の課題解決にも役立つ内容だと考えている」と述べたように、日本にも共通する課題だといえる。

もう1つの課題として挙げるのは気候変動。これを緩和させるために、郊外で農業から得られるバイオマス、太陽光、風力といった再生可能エネルギーを生み出し、郊外をエネルギー基地にするソリューションも求めている。

IBAチューリンゲンは、このルーラル・アーバンの関係をテーマとし、次の目標を掲げている。
・地域のリソースを発見し、強化する
・都市と田舎の新しい関係を創出する
・取り組みに参画する人々を前向きに支援し、支援する
・トレンドを指示するアイデアを認め、相互の連携を図る
・特別なプロジェクトの後押しと建設実施

また、広域な州全体をカバーすることはできないため、いくつかのエリアを選んでいる。プランニング・テーマは次の4つ。
・生活の場にふさわしい活気溢れる街区
・サステナブルな村落エリアの形成
・生産力のあるランドスケープ（林業、農業）
・資産価値のあるランドスケープ（〈チューリンゲンの森〉の意味を学ぶ。エネルギーだけでなく、美しくなければいけない）

これらを実施することでチューリンゲンの進化につながるのではないか考えている。「もちろん、IBAによって素晴らしい建物もつくることで、インパクトを持ち得ると思う」と語った。

また、IBA事業のクライテリアは下記の通り。
・プログラムとしてのポイント
・国際的な位置づけ
・地域としてのアイデンティティ
・イノベーションとその意志
・プランニングと実現の風土・文化
・事業性と持続可能性
・福祉と適正化

すべてがIBAのプロジェクトになるわけではなく、素晴らしいもの、優れたものだけが残る。他の地域へも応用可能でなくてはならない。さらにもう1つ重要なこととして「素晴らしいプランニングと実験の文化を

- create regional circuits
- establish new partnerships
- challenge existing standards

As main fields of intervention it proposes 4 themes:
- livable quarters
- sustainable villages
- productive landscapes
- landscapes for experiences

IBA Thüringen addresses not only municipalities, it also addresses individuals, societies, associations, etc., and invites them to submit project proposals. Proposals must not necessarily have constructions as their goal; every other kind of project that could provide solutions (IT-solutions, organizational solutions, etc.) is welcome as well.

However, visible results are to be presented in 2019 (intermediate results) and 2023 (final results).

In order to guarantee a high level of quality, the IBA will select the most promising projects and offer to these "IBA-candidates" expert advice for detailed design and implementation. Upon final evaluation according to IBA-criteria in 2023, successful projects will be designated "IBA-Project."

The selected projects will receive priority for financial support from the existing urban development promotion programs.

The first call for projects attracted some 250 project proposals. Almost 50 percent were introduced by private initiatives, so we can also understand this particular IBA as an example for Machidukuri.

The IBA-candidates selected from the first call for project proposals will be presented on September 30, and at the conference an initial overview of typology and content of the IBA candidates will be presented.

表1/Tab.1　IBAイノベーション / IBA innovation measure

Basic ideas of IBA	IBAの基本的な考え方
1. Identify actual and future challenges	－現実と未来への挑戦を確認する
2. Mobilise national and international top professionals in order to find new, innovative solutions to new problems	－国内、外のトップクラスの専門家を募り、新しい課題の解決策を見付ける
3. Implement best projects	－優れた提案を実際に建設する
4. Test and evaluate customer acceptance	－建設した建物で検証を行い居住者の満足度を評価する
5. Promulgate solutions	－解決策の普及を図る

図1　IBAシュトットガルト・ヴァイセンホフジードルング
Fig.1　IBA Stuttgart / Weissenhofsiedlung, 1927

もたらすものでなければならない」と語った。計画を立て、透明なプロセスにして人々に計画段階から見えるようにすることで、地域の都市開発の文化を人々の間で育てていきたいのだ。「見た目が素敵なだけでなく数十年続くものでなければいけないし、人々の幸せや適切性について配慮されたものでなければいけない」とも主張した。

■注目度の高い、開かれたプロジェクト

IBAプロジェクトは、今は自治体、企業、財団、協会、大学、個人、誰でも参加できる。法人、個人問わずいいアイデアがあれば参加することができる開かれたプロジェクトなのだ。

ただし、IBAプロジェクトに認定されるには、一定のロードマップに則ってステップを踏まなければいけない。

第1のステップは「プロジェクトの募集」。つまり「プロジェクト概要書」を出せばよい。この段階ではアイデアだけで構わない。そのあとに「1次評価」がある。そこで選ばれたものが「IBA候補者」となる。(表2)第2のステップとしてIBA委員会に「プロジェクト実施計画書」を提出する。ここで2次評価が行なわれる。「実施してもいい」と評価されたプロジェクトだけが第3のステップに進み、「プロジェクトの実施」となる。ほんとうによいものと評価されたプロジェクトだけが「IBAラベル」を付けることができる仕組みだ。

そして2019年には中間報告展覧会が開かれ、2023年には最終報告展覧会が予定されており、そこで一般公開される流れだ。

参加者にはさまざまなメリットがある。もっとも興味深いのは「自分のアイデアで社会の関心を引くこと」だろう。自らマーケティングしなくても、IBAラベルを得ることで注目を浴びることができる。プロジェクトの実施にあたってはIBA側から支援も受けられる。また、IBAプロジェクトからいろんなことを学び合い、ネットワークを形成するうえでの支援も準備されているし、候補プロジェクトが補助金や助成金を獲得するための支援も受けられる。プロジェクトの舞台となる各自治体に対しては、コミュニティ体制やインフラを改善するための援助もIBAが行なう。つまり資金的にも質的にも、あるいは最終的に成功までも手にすることができる。それがIBAプロジェクトに参加することのメリットだ。

■広くアイデアを募り、厳しく選ぶ

「一次提案募集の結果は248件だった」と明かす。このうち90%が州内から、10％は州外からだった。「応募者の50%が民間から提案ですごく驚いた」とシュトレープ氏。ただし、248件のうち、1次評価で16件まで絞ったという。

「落胆するかもしれないが、今後も二次、三次と提案募集の期間が設けられるはず。そうすることでよりよい、いろいろな提案が扱えることになる」と語った。

現段階のIBA候補プロジェクト16件のうち、「移民の社会的統合」を扱っているものがあるそうだ。前述のように、ドイツでは移民の市民化を目指した社会的統合や融和が求められている。そのプロジェクトは、教育機関を用いることで新たにドイツに来た移民がきちんと定住できるように考えられているという。

そのほかは、新しいハウジングの開発や郊外らしい大々的なコミュニティをつくる提案もある。足を向けなくなった教会の新しい利用構想や排水(汚水)のエネルギー利用と農業利用という提案もなされている。統合的な地域開発計画や分散型エネルギー生産の組織化も残っているそうだ。

候補プロジェクトの内容を聞くと、日本の課題を先取りしているものもあり、実に興味深かった。都市や建築の発展に沿って、あるいは社会背景によってトピックを選び、広くアイデアを募るIBAは、「サステナ

図2　IBAインターバウ(西ベルリン)
Fig.2　IBA Interbau, Berlin

図3　IBAエムシャーパーク
Fig.3　IBA EmscherPark, NRW Ruhr region
🟩 Green zone
🟦 River Water / Emscher
🟠🟡🔴🔵 different projects in the park, about 100

エムシャーパークの旧工業施設の再利用事例
New building built in the existing coal mining facility

ブル社会のまちづくり」を考えるうえで大きなヒントを与えてくれそうだ。「私自身もどういう結果になるか興味がある。どうぞ皆さんも 2023年に見に来てほしい」と締めくくった。

表2　IBAチューリンゲン・プロジェクトの進め方
Tab. 2 Project process of the IBA Thüringen project

図4　IBAフュルストピュクラーランド
Fig.4 IBA Fuerstpuklerland, former open-cast coal mining

図5　IBAハンブルグ -- ウォーターハウス
Fig. 5 IBA Hamburg – Water house, a summer resort

IBAハンブルグ -- 市内の"職・住"近接施設（流入民向け）
A facility for integration of foreign workers

IBAハンブルグ 2013 IBA Hamburg
移民の社会的インテグレーションを考えた職・住統合
Integration of housing and work: new chances for integration of immigrants

IBAハンブルグ --河に浮くリゾートハウス
Water house which moves in response with the water hight

チューリンゲン州の位置
Location of Thüringen state

チューリンゲン州データ / Data of Thüringen State

面積: 16,171 km² / Area 16,171 km²
人口: 216万人 / Population 2,160,000
密度: 350人 /ha / Density 350 persons per 1 ha

主要都市: Major Cities
ドレスデン Dresden
ワイマール Weimar
イエーナ　Jena
エアフルト Erfurt
アイゼナハ Eisenach

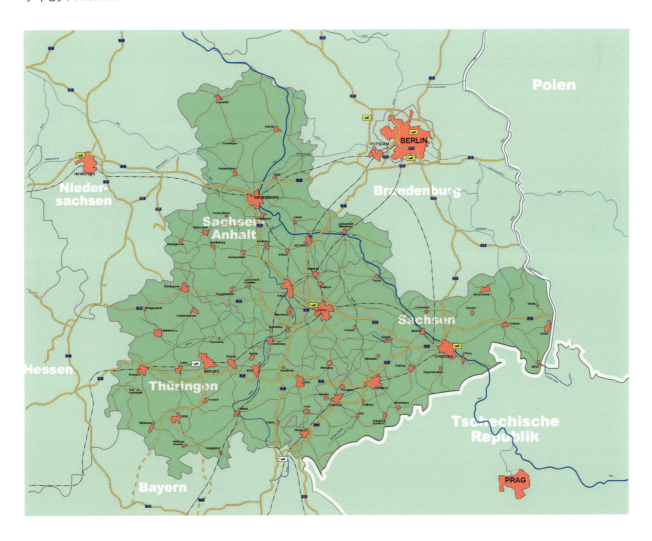

2-3 日本の団地再生・まちづくりの現状と課題
Recent Housing Refurbishment Projects in Japan

奥茂 賢仁
OKUMO, Kenji

■団地に必要なのは適切な建物改修と世代交代

　団地計画や住宅設計に従事し、一般社団法人団地再生支援協会のプロジェクト部会副部会長も務める奥茂氏は、「使い続けられる住宅づくり」という主題に対して、いま国内で進められている「団地再生プロジェクト」と団地再生支援協会で推進している「プロジェクト評価システムの概要」について紹介した。

　「築40年を経た高経年団地では、建物と設備の老朽化に加え、著しい高齢化の進展が社会問題となっているなか、住宅・福祉の連携方策が模索されている」と切り出し、中古住宅の更新・流通の停滞に伴う空き家も発生していると指摘する。これは住宅地のサステナビリティを確保していくうえで、きわめて重要な問題だ。

　その一方で、長い年月を経るなかで、団地には緑豊かで潤いある環境が形成されている。「適切な建物改修と世代交代ができれば、低コストで豊かな居住空間の創出・持続が可能である」と述べた。

　老朽化団地に内在するこれらの問題や特性は、近年ますます拡大、重層化しつつあるものの、一方で全国に団地再生プロジェクトが萌芽しはじめている。集合住宅ストックの現状や課題に即したさまざまな試みが行なわれるなか、それらのいくつかのプロジェクトの数例を紹介し、団地再生支援協会が試行研究として進めている「団地再生プロジェクト評価システム」の概要を提示した。

■日本の集合住宅団地の概況と課題

　昭和40年代から郊外に大量に供給された団地の総量は約2000万戸に達している。これには、公営・公団住宅などの「公共賃貸住宅団地」（340万戸）や、公的分譲を含む「民間分譲住宅団地」（571万戸）、民間賃貸マンションなどの「民間賃貸住宅」（1100万戸）に大別される。このうち、公共賃貸住宅団地に限ると、全ストック340万戸のうち、公営住宅220万戸、URがもっているものは75万戸、そのほかが44万戸である。重要なのは築30年以上の老朽ストックが半数以上となっている点。（図1）

　しかしこれらは日本社会のセーフティネットの1つとなっており、中長期計画に基づく計画的な再生が不可欠である。

　公営住宅団地には、築30年以上の老朽ストックが115万戸ある。トータルストックは219万戸である。「耐用年数が仮に70年とすると毎年3万1000戸建て替えないと維持できない。しかし、現実的にそれは無理なので、いかに長く使い続けるかがテーマとなっている」と解説する。

　一方、分譲住宅団地は、571万戸のうち築30年以上を経過したストックは106万戸となる。公的分譲は郊外立地が多く、老朽化した団地は大規模修繕工事を行なってリフレッシュすることが進められている。ただし、住民の合意形成が非常に難しく、建て替えに至るプロジェクトが非常に少ないのが実情。実施中のものを含めても再生事例は185件（平成23年）にとどまっている。

■「団地再生プロジェクト評価システム」

　奥茂氏は、次に団地再生プロジェクトの評価システムを紹介した。それぞれの団地再生活動の実態を把握し、整理・評価し情報の共有化を図ることは、全国的な団地再生を推進するうえで非常に有効だ。しか

■ Introduction

In Japanese old housing complexes which were developed about 40 years ago, the high-speed ageing of the residents became a social issue. This issue connected with deterioration of building structure and home appliances together are recently claiming more integrated policy of housing and welfare. Regarding the location of housings complex, "vacant room" problem is also the task to solve for housing corporations, such as UR, Housing Supply Corp. etc.

The amount of houses "valuable to refurbish" from those housing corporations have built last 40 years are estimated 2,4 million units in total. This is the subject to "Reuse/Revitalization of Housing Stock" measures of MoC policy since around 1995. (Fig1)

図1　公共住宅ストック形成の推移と現状
Fig.1　Development and present situation of "public housing"
注：公共住宅とは公団、公社、公営各住宅を合わせたもの
築後30年以上　1,150,000戸　older than 30 years 1,150 mill.
築後35年以上　806,000戸　older than 35 years 0.806 mill.

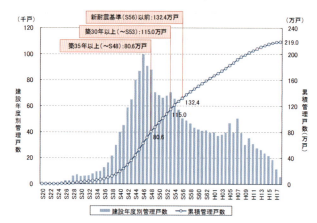

We, the first research association for housing refurbishment in Japan, have been working on the concept of the project in collaboration with several project initiatives since more than 15 years. Following is that we, Housing Refurbishment Association of Japan (HRA-J), are proposing for further working on this issue.

図2　団地再生プロジェクトの目指す3つの再生
Fig.2　Three tasks of housing refurbishment project

■ renewal, renovation of built environment
■ vitalization of dwellers' community
■ renewal of housing management system

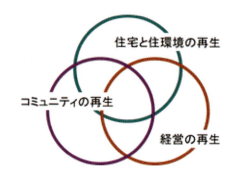

し、日本の団地はさまざまな戸数や規模が入り混じっている。そこで団地再生支援協会では、個々の状況・課題認識と解決策の整理に視座をおきつつ、"個"と"全体"を相対化し、今後のプロジェクトの進め方を導き出せるような仕組み「団地再生プロジェクト評価システム」の構築を進めている。

この評価システムでどのような効果が得られるのか。想定される効果として、奥茂氏が挙げたのは次の3点だ。
（1）先進的プロジェクトの情報を適切に流布すること
（2）個々のプロジェクトの状況、課題を計る「ものさし」をつくること
（3）その結果として個々のプロジェクトの進め方を求めるための仕組みづくり

団地再生プロジェクト評価システムをさらに推し進めて、いずれは団地再生促進のための制度創設、およびそのためのファイナンスとの連動も想定していると言う。

■先進的な3つの"流れ"

ふだん私たちはついつい「団地」と一括りにしがちだが、あらためて考えると、団地には賃貸－分譲、公共－民間、大規模－小規模、都市部－郊外部など、実はさまざまな違いがあることがわかる。

「数多くのプロジェクトがあるなか、相対化するために多種多様な再生プロジェクトを俯瞰してみると、そこに3つの先進的な"流れ"が見出せる」と語った。

1つめは、「居住環境の長寿命化」に重点を置いたプロジェクト。「ストック活用政策が功を奏し、萌芽的なプロジェクトが多数生まれており、住戸・住棟・屋外環境のさまざまな再生が試みられている」が、その一方で「改修の法的枠組みの未整備や合意形成、資金調達の困難さによって、所期の目的が未達成なプロジェクトもある」という現実もある。

ただし、住環境（土地・建物）の長寿命化を推し進めることは、人々が安心して暮らすための生活安定化の基礎となる。そのため、事業化の方策や法的取り組みなどに優れた方策を評価する必要がある。

2つめは「総合的団地再生のモデル」と見なされるプロジェクトである。すでに計画目標を達成したプロジェクトが誕生しているが、それらは見方を変えると「団地再生という事業が抱える多くの問題を解決した結果であると見なすことができる」と指摘。先進性の高い方策の結集によって、目標が達成されたと評価する必要がある。

3つめは、今回の主題にも通じる「サスティナブルな社会づくりのプロジェクト」だ。持続可能な社会の実現に立脚したプロジェクトとして、超高齢社会への対応、新たなエネルギーシステムの導入などによって、団地の枠を超え"まち"や"地域"で成立しているものを指す。つまり、きわめて明確に地域再生を誘導しているものがこれにあたる。

■多様な連携が必要な団地再生プロジェクト

団地再生プロジェクトの目標はどこにあるのか。奥茂氏は「住宅と住環境の再生」「コミュニティの再生」「ハウジング経営の再生」という「3つの再生」の組み合わせによって、実際の再生活動は動いていると見ている。（図2）

この視点に基づき、個別の団地再生活動を捉えるため、次の3つの条件を把握・整理する軸としている。それを「プロジェクトの環境」「プロジェクトの目標」「プロジェクトの体制」とした。

「プロジェクトの環境」は、そのプロジェクトがどのような条件下で計画されているかについて、立地条件、空間条件、運営管理の条件、入居状況により分類、整理する。

「プロジェクトの目標」は、そのプロジェクトがどのような目標に基づいているかについて、住環境の再生、コミュニティの再生、団地経営の

■ What we need for supporting housing refurbishment projects in Japan?

Overviewing Housing Estates/Complex to be Refurbished

In 1960's, housing complexes were supplied in suburban areas, the era of mass-housing supply. They are public rental housing, private housing and private rental housing types, about 20 million units in total. (More than the half of public housing and about 60 percent of HUD housing are over 30 years-old and antiquated stock. The private housing complexes, which were built by public sector, passed 40 years already. This part, about 140,000 units in total, are will be of a value to reuse thorough renovation and appliance replacement.

In contrast to this huge volume of housing refurbishment subjects, there are only several projects which have been brought to a successful result. Among 185 project cases we summarized their data in 2011, we see very complex problematics as regard to planning, design, construction and operation of the housings.

Housing Project Assessment System Study-Three Types

Our heuristics observation of refurbishment projects brought out three different project types:

Type-1: Sustainability-conscious inhabitants and well-market-oriented management are forming that project management. The initiatives spend good time and energy to set-up the projects.

Type-2: Inhabitants are interested for improvement of immediate living environment such as interior finishing and home appliances, and for the cost they have to pay for it.

Type-3: Housing community people are willing to spend their life there further on. The people are looking for solutions via usual media and personal contacts.

Characteristics of these project types:
The housing refurbishment "movement", in terms of its direction and necessary time, will be described in relation to following condition:

	Location of Project	Generation	Life Style
Type-1	Urban and Suburban and Sub-urban	Young & Senior Family	Mixed and active
Type-2	Sub-urban and Urban	Middle-age Family and Senior	Separately, individually
Type-3	Sub-Urban relatively small unit	Senior, Middle-age and Senior	Corporate

These projects are also tightly connected to municipal policies of:
   a. Energy Reform
   b. Welfare system for the aged people and young mothers
   c. Corporative management with municipality
• Town management

Housing Refurbishment Project Assessment System

A Data-Base System was developed several years ago which will show the condition and situation of individual project at a glance. The data are:

1.Project outline data – Housing Estate Condition
Name of housing estate and address
Start year of the development
Developer and management organization
Area (ha) and population
Number of dwelling units
2.Present situation and problems
In dwelling unit level

再生により分類、整理する。
　「プロジェクトの体制」は、プロジェクトの推進体制が備えるべき4つの機能（プロジェクト・マネジメント、資金や技術の調達、自治体からの支援、合意形成を図るシステム）に分類し、運営体制を整理する。
　こうした上記の客観的な情報に加えて、個々のプロジェクト固有のテーマの評価軸として5つを挙げた。

（1）エネルギーの合理化
（2）少子高齢社会への対応
（3）防災・防犯性の向上
（4）地域社会への貢献・調和
（5）サステナビリティの確保

　これら5つの評価軸を設定したうえで、各プロジェクトが特化して取り組むテーマが先進性や妥当性、合意性を有するかについて客観的に整理するとした。
　さらに、各地で進められる再生プロジェクトから事例を抽出し、各々の位置づけを明らかにする試みを通じ、その情報をデータシート化した「プロジェクデータ」を作成。（図3）
会場ではこれを簡略化したものを投影した。概要のみ抜粋する。

○プロジェクト01　洋光台団地（神奈川県横浜市）
　209ha、5573戸の大規模団地。UR都市機構と横浜市によって開発。分譲と賃貸が混在している。「郊外住宅地活性化のモデル」として、まちづくり協議会が組織され、今後のまちづくりビジョン検討が始まっている。専門家によるアドバイザー会議が組織され、まちのプロモーション活動にも着手している。

○プロジェクト02　若葉台団地（神奈川県横浜市）
　神奈川県住宅供給公社が1979年から開発。90ha、6304戸とこちらも大規模な団地。同公社ではセンター地区の魅力向上や、高齢者への生活支援、若年世帯への訴求力向上、エネルギー使用の合理化、防災性の向上など、団地再生に対して総合的な取り組みを進めている。

○プロジェクト03　観月橋団地（京都市）
　UR都市機構が1962年から開発。3.3ha、540戸の小規模団地。集約型の建て替え事業ともに、建て替えによって生じる余剰地を新たなまちづくりに活用する予定である。また、既存の住宅をリノベーションすることによって、新たな需要層の開拓にも取り組んでいる。

■これからの団地再生まちづくり

　奥茂氏は、「日本の住宅産業は、高度経済成長期における大量な住宅供給にはじまるスケルトン・インフィル型の産業化が発展した。これは国際的にも高い水準にあり、団地再生の促進に貢献できる。しかし「団地再生」が「まちづくり」へと拡大している状況にあり、少子高齢社会への対応、エネルギー変革などとの連携が不可欠である。」と述べ、団地再生支援協会のこれからの活動の重点を3つ取上げた。

1. 団地再生・まちづくりに関する知識・情報の集積センターとしての機能を充実させ、希望者の要望に的確に対応する情報提供ができるようにする。
2. こうしたセンター的活動を通じて、プロジェクト主体者および各種課題に関わる専門家との連携体制の整備を進める。

In house and surrounding's level
In housing estate level
3. Refurbishment project data
　Project organization
　Program outline

We developed a Data-Base Sheet of page layout like this:

In this sheet, the data 1, 2, 3 are put together in each field.
As for the data not able to express with text but with picture some fields are prepared within the sheet.

■ Tasks of Housing Refurbishment Association(HRA) of Japan

Knowledge Centre
　Concerning promotion of housing refurbishment in Japan, HRA has a unique position to provide project initiatives with the knowledge they need for planning, design, construction of projects. The knowledge is extensive. Therefore it should be arranged for efficient dissemination.

Knowledge Dissemination
　HRA disseminates relevant knowledge (information, seminar etc.) to everyone who are interested in housing refurbishment.

Project Assessment System Development
　HRA will initiate further development of the assessment system.
Procedure:
1. Collection of project examples – about 50~100 cases
2. Evaluation system development 1- purposes
3. Evaluation system developmet 2 – criteria system
4. Evaluation system developmet 3 – 1st system completion
5. Test run & improvement system

図3　団地再生支援協会と連携するプロジェクト事例
Fig.3　Project examples in collaboration with HRA in Tokyo & Osaka
🟥 Condominium development　　🟦 Public rental apartments
🟨 Rental & Condo- mixed housing

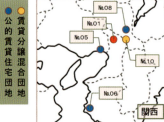
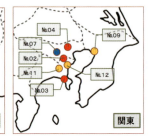

3.団地再生プロジェクトの評価システムについては、以上の支援協会体制の充実に合わせて、各課題ごとの評価の仕組みを構築する。最後に前年から始まる国土交通省「地域居住機能再生型大規模団地再生」事業を紹介し、状況に関する理解を求めた。(図4)

図4　国交省地域居住機能再生型大規模団地再生の考え方
Fig.4　MoC "Compact City" Development Program - Concept

The Left (Before) situation moves to the Right (After)
Whereas following measures will encourage the project:
・Bringing together residential buildings to the central area
・Up-grading of the left open spaces
・Efficient management of housing and town facilities and sevices

Project Data-01

洋光台団地／ YOUKOUDAI Housing Estate

○プロジェクトデータ／ Project Data
・所在／ Place: 神奈川県横浜市／ Isogo, Yokohama, Kanagawa Pref.
・開発年次／ Date of Development ; 1970〜
・開発主体／ Developer ; UR都市機構、横浜市／ UR ,Yokohama City
・開発規模／ Area ,Population: 209ha ,26,000 people
・住宅総数／ Number of dwelling units: 5,573 units
・種別戸数／ UR: 3,350 units, Public Housing: 730 units,
　　　　　　　High-rise bldg: 1,493 units / Low-rise bldg: 6,449 units

○現状と課題／ Present Conditions and Problems
・築40年を経過し、住民の少子高齢化の進展に合わせ、地域活力の浮揚、公共サービスの充実が必要である。／ Over 40 years, with the persistently declining birthrate and growth of the aging population, improvement of the town vitality and the public service are necessary.
・様々な住宅の混成による郊外立地の大規模団地であり、各住宅セクターだけでなく、活性化に向けた横断的な取り組みが必要となっている。／ So it is huge suburban housing complex, not only each developer but every town manger have to work on re-vitality of the town.

○再生への取り組み／ Tasks of Refurbishment project
・今後の郊外住宅地活性化のモデルとして、まちづくり協議会が組織され、今後のまちづくりビジョン検討が始められている。／ As a model of the future suburban town activation, a meeting made with the town was organized, and vision examination made with a town of the future is begun.
・専門家によるアドバイザー会議が組織され、まちづくりのプロモーション活動に着手されている。／ The adviser meeting by the expert was organized, and they are improving town promotion activities, and any other plans.

全体概要／ General Plan

Project Data-02

若葉台団地／WAKABADAI Housing Estate

○プロジェクトデータ／Project Data
- 所在／Place: 神奈川県横浜市／Asahi, Yokohama, Kanagawa Pref.
- 開発年次／Date of Development ; 1979〜
- 開発主体／Developer：神奈川県住宅供給公社／Kanagawa Pref. - Housing Supply Corp.(KHSC)
- 開発規模／Area ,Population; 90ha ,17,000 people
- 住宅総数／Number of dwelling units; 6,304 units
- 種別戸数／KHSC rental housing: 792 / 5,186(Estates) , 326(Rental Housings for Elders), 92(Nursing Home)

○現状と課題／Present Conditions and Problems
- 築35年を経過し、空き家率を抑制するために、街の魅力の維持・向上が必要である。／Over 35 years, improvement of the town attractiveness are necessary to restrain vacancy rate.
- 賃貸、分譲間での世帯の移動、域外からの世帯の導入による、住まいの循環が必要となっている。／So it is necessary to improve circulation of household among rental housing and estate housing and other area.

○再生への取り組み／Tasks of Refurbishment
- 公社ではセンター地区の魅力向上や、高齢者への生活支援、若年世帯への訴求力向上、エネルギー使用の合理化、防災性の向上等、幅広い団地再生に総合的に取り組んでいる。／KJK are improving the attractiveness of town-center, life support plan for elders, advertising appeal to young people, rationalization of energy use, disaster prevention, and any other plans.

- 再生のテーマとして「地産地消」が掲げられている。／KJK are dvocating「Local production and consumption」as the theme of refurbishment (energy, foods, service, human...).

全体概要／General Plan

Project Data-03

観月橋団地／KANGETSUKYO housing estate

○プロジェクトデータ／Project Data
- 所在／Place: 京都府京都市伏見区／Fushimi, Kyoto, Kyoto Pref.
- 開発年次／Date of Development ; 1962〜
- 開発主体／Developer；UR都市機構／UR
- 開発規模／Area ,Population; 3.3ha ,1,200(people)
- 住宅総数／Amounts of Units; 540(houses)
- 種別戸数／Each Units; 540(UR Rental Housings)

○現状と課題／Present Condition and Problems
- 築50年を経過し、建替え、リノベーション、用途転換含めた総合的な街の再生が必要である。／Over 50 years, improvement of the town refurbishment are necessary, by the renewal, the renovation, switching of the land use.
- 高齢者の居住環境を整備すると共に、若年層の導入による街の活力の向上が求められる。／Improvement of the residence environment of the elderly person, and improvement of the vitality of the town by the introduction of young people is necessary.

○再生への取り組み／Tasks of Refurbishment
- UR賃貸住宅の集約建替と共に、既存住宅のリノベーションにより、新たな需要層の開拓に取り組んでいる。／UR are working on new-customer acquisition, through the collection by the rebuilding, and renovation of old housings.
- 建替えによる余剰地を新たな街づくりに活用する予定である。／UR are going

全体概要／General Plan

2-4 「箱」の産業から「場」の産業へ―まちづくり産業の転換
From "Box"-making to "Place"-making industry – Change of Machizukuri Industry In Japan

松村 秀一
MATSUMURA, Shuichi

■「箱」の産業から「場」の産業へ

建築構法や建築生産を専門分野とする松村氏は、日本における建築または都市にかかわる産業が転換期を迎えている今、どのような産業的転換を果たしていかなければならないのかという見地から、自身の考えを示した。

まず、日本では40年以上続いた大きな新築市場があると指摘。人口1000人当たりの住宅着工戸数を表したグラフを見ても、日本は圧倒的に高い推移だった。「世界史的に見てもきわめて異常な状態だった」と述べた。（図1）

この市場環境のおかげで、工業化や都市化に関連する様々な技術の開発と適用を進めることができた。日本には100の大学で建築学科に相当するものがあり、年間1万4000人を毎年輩出している。さらに、これまではそれをすべて吸収できるような産業規模を維持できたのは、高度成長期以降、延々と新しい建物をつくり続けてきた結果だ。

しかし、考えなければならないのは、日本全体が建物、新築に非常に大きな投資をしてきたこと。6000万戸を超える住宅があり、人口で割ると0.48戸、つまり2人に1戸の計算となる。これは先進国のなかでも最高水準で、日本の住宅ストックはあり余っている。空き家の戸数は800万を超え、決して古いものではないのに空き家率は13.5％に達している。

こうした市場環境の変化が建築産業に転換を促している。対象とする事柄は変わらざるを得ない。「それを一言で表せば『箱の産業から場の産業へ』ということだ」と語った。

箱の産業とは新しい箱（建物）をきちんと生産して供給することにのみ集中する産業のことで。この面では日本の建築産業は非常に優れている。しかし、先ほどのデータに表れているように、新しい時代はもはや新しい箱をこれ以上必要としていない。それよりも、すでにある空間をより豊かでいきいきとした人々の生活の「場」に変えていくことが求められる。「場の産業とはそうした需要に応える産業だ」と述べた。

「何か活動をしたいと思っている人」と「豊かに存在している空間資源」をどう結び付けるかを考えると、新しいタイプの仕事が生まれてくる。すでに新しい産業の萌芽が、この5年ほどの間に日本の各地に現れているという。小さな点が面に広がっていくタイプの先端的なまちづくりに取り組むのは、30代半ばから40代半ばが主流だ。

「個別の建物のリノベーションを積み重ねながら、エリアを変えていく仕事をしている人が多い」と分析する。（図2）

■建物のリノベーションから地域再生へ

場の産業を形成するために不可欠な事柄として「新しい仕事の基本要素」を5つ挙げた。
（1）生活から発想する／利用の構想力を導入する
（2）余剰空間を「空間資源」と捉え、発見する
（3）空間資源の短所を補い長所を伸ばす
（4）「場」化する＝新しいアクティビティを導入する
（5）人々と場を出会わせる

松村氏は事例を交えつつ、順を追って説明していった。いちばん重要なことは「生活から発想すること」。箱の産業は生活のことも考えているけれど、箱を届けることが最大の目的。しかし「そこで何をしたいの

■ Major Change of the Building Market in Japan

In Japan, the building industry had much enjoyed the extraordinary growth in the new construction market for a period of about 40 years beginning in the middle of the 1960s. Thanks to the long-lasting and large new construction market, various technical developments had occurred and new technologies applied to industrialization and urbanization. (Fig.1)

However, the new construction market has declined during the past five years. As a result of vigorous activity in durable construction, we now have too many buildings throughout the country. For example, there were 60.6 million housing units in 2013.

This number exceeds the number of households by 8 million. Currently, there are more than 8 million vacant housing units and the vacancy rate is at 13.5 percent, although these units are not that old. Such a change in market conditions gives impetus to the building industry to make some changes.

■ Changes Needed in the Building Industry

In short, the changes needed in the building industry can be described as "from the Box-Delivery Industry to the Place-Making Industry." The Box-Delivery Industry means the industry that concentrates only on new buildings. From this point of view, the building industry in Japan is well trained.

But we do not need any more boxes, or buildings. Instead, we have to make already-existing spaces more comfortable and life enriching. The Place-Making Industry refers to the industry that can respond to such market demands. It is clear that these two industries are different. The essential goals of new professionals in the Place-Making Industry are

1. to show users how to use existing spaces creatively
2. to regard vacant spaces as our rich resources for better living
3. to develop the strong points of the spaces' resources as well as to strengthen their weak points
4. to introduce new activities people can enjoy in existing boxes
5. to encourage people to find a place they can utilize

(Fig.2)

■ From Building Renovation to Area Revitalization

We can see new phenomena of the industry's change during the past five years in a number of regions across Japan.

What is interesting about these phenomena is that they initiated new professional activities with the renovation of each building in an area, and the continuous renovations revitalized that area.

Typical examples can be seen in Central Tokyo, Nagano, Okayama, Kita-kyushu, and so on. (Fig.3, 4, 5, 6)

か、どう使いたいのか」がなければ場の産業のプロジェクトが発生しない。空き家を住めるようにと仮に雨漏りを直しても、生活がなければ何も動かないのと同じだ。生活者自身に眠っている利用の構想を引き出し、「利用の構想力を組織化していくアプローチ」が必要となる。

そして、空いている箱を「場」にするためには、新たなアクティビティを導入することも重要だ。「簡単にいうと、われわれの産業はコンテンツ産業になっていくと思う」と主張した。

1つの例として、統廃合で廃校になった旧練成中学校を現代アートの発信拠点として活用している東京都千代田区の「3331 Arts Chiyoda」を挙げた。過去4年ほど運営されているが、年間80万人が訪れている。空いていた箱を、アートセンターという生活の場に変えたのはアーティストの中村政人氏。東京の秋葉原は世界に向けて発信力のあるサブカルチャーの発信地になると見抜いていたところ、ちょうど千代田区が旧練成中学校を定期借家方式で事業者を募った。中村氏は事業会社をつくり、見事に借りることができた。今はギャラリーやアート系の事務所からの家賃収入で運営している。また、東日本大震災の直後に津波で流された写真を洗って修復する作業を行ない、震災復興展を開催し、時には味噌仕込み教室も開く。

こういう場をつくっているのは運営会社のコマンド A。どういうアクティビティをいつどこでやるのかを決めている。その結果、常に生き生きとした活動の場になっている。「従来の建築産業はまったくかかわってこなかった分野だが、今後はこうしたことが中心になっていくのではないか」と述べた。

■空き家ではなく「空間資源」と見る

空きスペースを豊かな空間資源として捉え直すことも大切だ。空き家問題と聞くとネガティブなイメージを抱くが、逆に「空間資源」ととらえれば可能性は広がる。この観点から活動しているのが北九州家守舎というエリアマネジメント会社だ。オーナーの協力を得て、地域で空いている建物の情報をとにかく集める。それをもとに、個々のオーナーへ適切にアドバイスしてテナントを見つけてくる、あるいは、パン屋を開業したい人が現れたら「どこが適切なのか」と彼らは考える。あらかじめつかんでおいた空間資源の情報を活かして、プロパティマネジメントとエリアマネジメントをつなげていく仕事だ。そこには「このエリアをどうすれば、まち全体の賃料が上がっていくのか」を常に考えている。

次に同じく北九州の事例にふれた。松村氏が副理事長を務める一般社団法人HEAD研究会は、3年ほど前に北九州市の支援を得て小倉北区魚町(うおまち)で「リノベーションスクール」をはじめた。過去3年間に7回開き、全国の自治体の職員、まちづくり系のNPO、設計事務所、不動産会社、学生など延べ数百人が卒業。毎回4日間にわたり、10人ずつの異業種混合チームを10チームつくる。各チームには課題として実際に魚町にあるビルを与える。各ビルのオーナーには「もしいい事業提案が出たら採用を考えてください」と事前にお願いしてる。(図3、4)

4日間、10人が事業収支も含めて徹夜で提案をまとめて最後に公開プレゼンをするが、オーナーも全員来てもらう。日本の特性として、非常に小さな土地に小さなビルをもつ小さなオーナーによってまちはほぼ構成されているので、小さなビルのオーナーをどう巻き込んでいくかはきわめて重要だからだ。オーナーは各地から来た若者たちが自分のビルのために必死に事業提案をつくるので感激する。投資効果も得られるいいプレゼンとオーナーが認めて、これまで15棟が事業化にこぎ着けた。

このスクールの指導者として、いくつかの萌芽的な先進プロジェクトを実現した人たちが付き添っている。するとソーシャルネットワークでつながって、どこでどういうリノベーションが行なわれているか、今はみんなが知っているという副産物も生まれている。

図1　人口1000人当たり住宅着工数推移
Fig. 1 Development of housing production: 1950 - 2008
■ Japan ■ France ■ UK ■ Germany ■ USA
(Units /1,000 people.)

図2　建築産業に必要な変化
Fig. 2 Industry changes required by market change

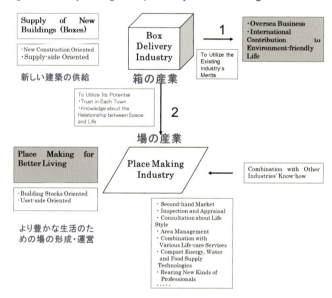

図3　北九州市の空き家を記した「小倉ポテンシャル map」（北九州家守舎・九州工業大学徳田研究室作成）
Fig. 3 Map of vacant houses in Kokura district, Kita-Kyushu

また、築50年の団地という空間資源の可能性を示す再生例として東京都日野市にある多摩平団地の「たまむすびテラス」を挙げた。URが古い団地5棟の運営を事業者に任せて見事に成功。新しいまち開きの際、50年前に住んでいた人たちを招くと懐かしんで集まってきて、新しい住民との交流も生まれた。こうしたアクティビティは運営会社が仕掛けたものだ。「学園祭のノリのようだが、これが大変リアルな社会現象になっている」と語った。

岡山市の問屋町の事例も紹介した。RC2階建てのまったく同じ建物が等間隔で並んでいる不思議な景観のまちだが、流通が変わり廃業する問屋が増えて空き室が目立っていた。そこで地元出身の若者が「違うまちにしていこう」とジーンズショップやおいしいパン屋など新しくて少し尖ったテナントを入れるように仕掛けた。すると点が面に広がり、非常に注目されるエリアとなった。岡山駅からタクシーに乗ると2500円かかるので多少距離はあるが、坪当たりの賃料は岡山駅前と変わらなくなったという。リノベーションのまちづくりの世界では「行くべきまち」と見なされている。（図5）

■人と場の出会いをつくる

場をつくったとしても、外から人が訪ねてこなければ経済的に行きづまってしまう。そこで「人と場を出会わせる」ことが必要だ。SNSなどのバーチャルではなく、実際にまちなかで出会うおもしろい事例として長野市門前町を挙げた。

「自然発生的に生まれた動きなのだが、古くからある空き家にまったく新しい仕事をする人が、しかも県外からどんどん入居している」と説明する。

出会いの場をつくる中心的な役割を担うのは『長野・門前暮らしのすすめ』という冊子だ。アーティストの夫婦が長野市からの助成金をもとにつくっているもの。冊子で「長野市の門前町でこんな暮らしをはじめた人たちがいる」と紹介するだけでなく、ライフスタイルの感度が高い人の集まりそうなカフェに置いてある。「えっ、長野ってこういう暮らしできるの？」と興味をもったら、月に1回行なわれている「空き家見学会」というイベントに参加するという仕組みだ。（図6）

空き家展覧会は不動産仲介兼建設業を営む「マイルーム」が行なっている。市場には出ない空き家は、実は結構ある。それを家主のおばあさんに「あの家を貸してもいいですか？」と頼んで許可をもらってからつないでいる。かなり長いウェイティングリストがあると聞いているし、今どんどん人が集まってきていて、ぼろぼろの空き家に若い人たちが移り住んでいる。こういう人を「ダウンシフター」と呼ぶ。収入を下げても充実した時間を過ごしたい、思い通りの仕事をやりたいという人が東京を離れて、とても安い家賃で充実した日々を過ごしている。

さまざまな事例紹介を通じて、新たな産業の萌芽を感じることができた。ちなみに北九州市の魚町では、15棟が事業化したことで新たに300人もの雇用が生まれているという。

「まだまだ小さいけれど、こうしたことが各地で、しかも連動して起きている。これを新しい産業の種にして、仕事をつくっていければと思う」と語った。

図4　一般社団法人 HEAD研究会の「リノベーションスクール」最終プレゼン（写真提供：北九州家守舎）
Fig.4　Scene of the workshop "Renovation School"

図5　岡山問屋町に新たに埋め込まれた店舗や事務所
Fig.5　Renovation of the whole sale area in Okayama

図6　長野市門前町で定期的に行われる仲介イベント空き家巡り（写真提供：㈱マイルーム）
Fig.6　Monthly tour of vacant buildings in Nagano

2-5　ゲーラ2030まちづくりプロジェクトの仕組み
Gera 2030 Project – Integrated urban development project

R. ミラー
MILLER, Ramon

■ 東西ドイツ統合後の苦闘

　ゲーラ市の副市長として都市開発と環境整備を担当しているミラー氏は、ゲーラ市の現状と今進められている統合的まちづくりコンセプト（ICDC）について報告した。

　人口およそ9万5000人のゲーラ市は、チューリンゲン州第三の都市である。ドイツの中央部に位置し、有名なワイマール、イェーナ、ドレースデン、ライプチヒに連なる（図1）。ベルリンからは2時間ほど。この地域は、東西ドイツの統合後、首尾よく開発が進められた結果 GNPと収入が増え、失業率も低下するなど明るい将来計画をもつ都市が多い。しかし有力な都市が周辺に多いがゆえに、ゲーラ市が独立を保つことが難しいという。

　ゲーラ市は777年前の1237年に都市権を受けた。繊維業が盛んで、500年間ほどドイツ帝国の州都だった。19世紀初頭には紡績機が導入され、またドイツ国内でも早くに鉄道とつながった。1892年にはドイツで2番目となる路面電車が走り、1902年にはアールヌーボーの劇場が建てられた。旧東ドイツ時代はゲーラ市の東部でウランが採掘され、重要な都市だった。

　1959年以来、人口は常に10万人以上をキープし、最大人口は13万5000人。しかし、ドイツ統合後は一転する。州都ではなくなり、たった2年ほどで競争力のない繊維産業、機械製造業、軍需産業が縮退し、失業者は2万5000人に上った。2014年の人口は最盛期よりも 4万人も少ない。「ドイツ国内でもこれほどひどい人口減はない。空き家は1万5000戸もある」と言う。

　失業率は1994年が19%で2014年は9%。回復したように見えるが「多くの人たちがほかの都市に移ったうえでの数字で、1994年当時の失業者はすでに現役を退いており失業統計にカウントされないことも考えなければいけない」と語る。

■「野心的に前へ進む」という伝統

　1990年以降、何もしなかったわけではない。民間投資によるリテール業やサービス業の急成長があった。衰退したとはいえ伝統的産業はまだ残っているし、テキスタイル、機械製造、光学などの高度加工は競争力を保っている。1997～2007年には連邦政府による集約的投資が行なわれ、鉄道やLRT、道路、スポーツ施設、文化施設など総額5億5000万ユーロが投資された。1991年以降は、総額2億5000万ユーロのハウジングへの公的補助付き民間投資も行なわれた。中心市街地には再開発エリアがあり、新たな都市型公園もつくられ十分暮らすことができる。（図2）

　しかし、たとえば50年後などを考えると心許ないとミラー氏は言う。ゲーラ市は努力を続けているが、経済活動はドイツの他の地域に比べて依然弱い。高齢化も進み、2014年の平均年齢は45歳で2030年には50歳となる。若い世代の流出も止まらない。

　市の財政は、収入の少ない割に社会支出が多く、州の平均値を上回っている。つまり、経済活動が力不足のため、足を引っ張っている格好だ。

　「しかし、ゲーラ市にはこれまでの歴史が培ってきた『チェンジと発展』がある」と主張する。ゲーラ市は幾度もの壊滅（1450,1639,1686,1780,1945各年。戦災と火災）とその後の再

■ Struggle after the re-unification of Germany, 1990~2010

The city of Gera is – with 95.000 inhabitants - the third largest city in the State of Thüringen. The city is lies in Central Germany. It is surrounded by famous cities like Weimar, Jena, Dresden or Leipzig. The travelling distance to Berlin is only two hours. The Region is developing well since the Reunification of Germany in 1990: Rising GNP and income, falling unemployment rates with a positive perspective for the future. (Fig.1)

Gera takes part in this strives for better conditions, but has had a very heavy starting point. The industries that were pushed forward during the GDR period collapsed almost completely within a short time of only three years. This - and poor housing conditions - caused migration out of the city. Gera lost about one third of its population – 40.000. And the city grew older: the average age now is 45. Tendency: rising. The municipal budget is unbalanced with comparatively small income and social expenditure above the average of Thuringia. A result of the weak economy which severely reduces the capacities to take an active role – especially in the field of urban development.

But Gera has opportunities:

1. The public sector has invested in reconstruction and modernization of infrastructure: railway, major roads, monuments, schools and urban renewal in historical districts.

2. The private sector has benefited from these efforts and also invested in new companies, housing and retail.

3. It has a rich cultural heritage and still considers itself to be a city with rich cultural life. It is also a sporting city.

4. It has a strong tradition of ambition to push the city forward. Textile manufacturing started in the 16th century, industrialization in 1850, electrical trams are operating since 1892 etc.

Thus dissatisfaction with the present situation is widespread - unfortunately. (Fig.2)

■ In 1998 a master plan for the city development was passed by the city council.

It focused on a very extensive public investment program which was managed by the city administration. The master plan was financed by public funds – and Gera sold a housing company, together with the public hospital. This program was the main element of a strategy to stop further decline. That was achieved and the turning point reached. But sustainable and self-supporting development is not yet established.

Concerning this background, the city council decided a new „Integrated City Development Concept" (ICDC, in German: Integriertes Stadtentwicklungskonzept, ISEK). This decision contains new approaches (Tab.1)

1. The main actor of future urban development is not the public sector with funds and administration. The main actors should come from the civil society: inhabitants, private supporters from abroad, businessmen and multi-players

2. The main focus is not on public expenditure. It should be on the encouragement of private (financial) initiatives

3. The ICDC Gera should create a platform for a broad public discussion:
・What are the crucial issues for the future in Gera?
・Which are the main goals for the year 2030?
・Who will achieve those goals – and how?

建を経験している。この歴史が市民の生活の一部となっていて、困難を乗り越える自信がある。「野心的に先に進む」伝統が息づいているのだ。

■ 2年半かけて固めたコンセプト

1998年に都市開発マスタープランが議会を通過、公的資金によって進められ、ハウジング会社や医療施設経営組織に譲渡されてきたが、それが終わると曲がり角に立つ。

サステナブルで、自立性の高いまちづくりを進めるために、市議会は統合的まちづくりコンセプト(ICDC= Integrated City Development Concept)の設定を決定した。「ICDCは計画のためのツールだ。土地利用計画があるわけではない。インフォーマルなもので、市議会が決定した方針だ」とミラー氏は説明する。

この決定には次のアプローチが含まれている：
（1）広範な市民参画の実現
（2）社会計画と都市計画の統合
　（老若男女それぞれの社会的問題を都市計画に反映させる）
（3）ネットワーキングしつつ実施に移す試行錯誤を進める
（4）市議会が何をしているかを知らせる広報宣伝活動の強化

さらにICDCは「主役は市民コミュニティ」と考え、民間支援者、企業、マルチプレーヤーも引きこみ、民間イニシアティブ（投資）の促進も図る。次の4つがアプローチの要素だ。（表1）
①なぜか？＝総合的条件の設定
②どのように？＝未来発展の戦略とガイドライン
③何を？＝プロジェクトにおける公・私共同活動
④いつか？＝実施過程

「ICDCはうまくいかないだろうと言う人もいた。しかしうまくいった」と語るそのプロセスは、まず2012年にスタートアップのための公開議論を市議会会場で開催。市民全員が招待され、一般的な討議を経て、15のワーキンググループが発足する。

次に、グループごとに10〜15名の参加者が選ばれ、数回のワークショップを開く。6カ月を経てガイドラインを決定する。さらに、テーマ別のワークショップによって、既存プロジェクトの具体化および新規プロジェクトの作成を進める。ICDCの案（ドラフト）が公開され、パブリックコメントがインターネット上で行なわれる一方、まちづくりPRの一環としてフォトコンテストを実施。また、公共機関や大学研究者にICDC案の評価、意見を求めた。最後は市議会の諸委員会でICDCの集中議論を行ない、2014年5月に「ICDCゲーラ2030」事業は賛成多数で議会決定した。

2年半の間、このようなプロセスを経てICDCのコンセプトを固めていった。そのプロセスを支援したのは、自治体執行部の専門職、ICDCプロセスのマネージメント専門家（コンサルタント）、関連省庁の専門職、大学研究者、地元の専門家など。シュトレープ氏も参加したという。

■ 自立性の高いまちづくりを可能に

こうしてつくられた全体コンセプトのトピックは、将来のまちづくりを定義する戦略要素だ。「まちづくりは市民コミュニティの継続的参画により推進」「名誉職的関与をお願いする人、イニシアティブの関与を支援する」「地元、地域、国のレベル、国際のレベルのネットワーク化」「独自の強みを再認識し、拡大を図る」など。「私がここで今話しているのもゲーラのためのネットワーキング活動の一環なのだ」と笑う。

これらの真の目的は「一番良いソリューションを見つけることではなく、ゲーラ市でうまくいくベストソリューションを見つけること」。そのため、古くならない、フェアなまちをつくり、生涯学習機会を拡大し、生き生きとしたアーバンセンターと住みやすいまちをつくることを目指している。

開発されたゲーラ市のマスターコンセプトは、広い範囲の人々に提示

■ The start-up of the public discussion took place in January 2012 in the city hall.

It was an invitation to every citizen of Gera. After general introductions teams of about 10-15 participants were established and several workshops carried out. Within six months the overall concept defining the main issues and the general mission was created.

A special committee of the city council "polished" and "legalized" this ground concept. The first and very important step was taken. And despite of fears at the beginning of the process: there were no major dissents between the opinions of citizens, the local politicians, the administration and involved experts.

The main agreed strategic elements are:
- developing the city together by continuous participation of the civil society
- supporting the commitment of honorary offices and initiatives -networking on local, regional national and international platforms
- recognition and extension of inherent strong points

The main agreed fields of public-private action are:
- future-compliant working world and strong economy
- socially fair city with an active citizenship
- applied science knowledge and leading through education
- ivable city with a vivid urban center

■ The workshops – now structured by the fields of action

- Citizen's initiative continued development of the master concept of Gera within another period of six months. Goals and projects were discussed, defined and presented to a wider audience.

Thus feedback with a positive response encouraged further engagement in the ICDC-process. 12 lead projects form the basic framework for the city development until 2030. This framework consists of over fifty subprojects covering crucial issues in all agreed fields of action. The project managers are widespread and only a small part regarding city development and management is now financed and implemented by the city.

The entire process was managed by experienced urban planning consultants. The professional input was very helpful, but not the only source of ideas and engagement.

■ The concept draft was made

The concept was discussed by the city council and by almost all of its committees - and not only those concerned with urban planning and development. The ICDC Gera 2030 was finally passed by the council with an overwhelming majority in May 2014. (Fig.3)

Considering the general approach and its results, the City of Gera is now on a path that offers opportunities that can lead to sustainable and self-supporting development with adequate growth. The city is able to deal with the future tasks and the development.

され、ポジティブな反響を得た。そして「12のリーディングプロジェクト」が2030年目標達成のまちづくりの基本フレームワークとして組織された。

このリーディングプロジェクトは総数50のサブプロジェクトから構成され、一体となってアクション分野の必須の要件をカバーしている。各プロジェクトのマネージャーは広範な分野からの人々であり、市の執行部から資金を得ているのはごくわずかにすぎない。

具体的には、2030年までにゲーラ市再生の枠をつくる先導プロジェクト12件を設定。その他の35件のプロジェクトは市の執行部が支援する多様な体制で行なう。トータルで47のプロジェクトがスタートしている。

初期のICDCプロジェクトによって、すでに新たな産業エリアが開発された（チューリンゲン州開発公社や欧州地域開発金融資金による工業団地の実現）。公立学校を整備するマスタープランをつくり、家族にやさしいまちづくりの監査も行なった。

また、800haの森を「都市の森」としてレクリエーションにも使えるようなプランを作成中だ。これはNPO組織とエアフルト応用科学大学の協力によって評価する。

さらに、IBAチューリンゲン2023事業の1つに採択された「ゲーラ・シティセンター強化プロジェクト」もある。これは10年前に住宅が取り壊された中心市街地のオープンエリアについてPPPを念頭に国際コンペを開き、投資を含めてこの地域を再開発するもの。「これはゲーラ市として最大のプロジェクトだ。2023年にきちんと開発された地域になればいい」と語った。（図3）

全体のアプローチとこうした成果を概観すると、ゲーラ市は「サステナブルで、自立性の高いまちづくり」によって豊かな生活を提供できる立場になったと考えている。「いまゲーラではいろいろな人を海外から呼んでいるので、興味のある人はぜひ一度活動に参加してほしい」と結んだ。

図1　ゲーラ市の位置／チューリンゲン州の東、ベルリンに2時間
Fig. 1 Location of Gera city in Thüringen, 2 hrs's drive to Berlin

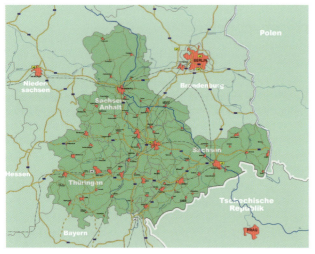

図2　まちの中心街—再開発エリア、新都市パークをのぞむ
Fig.2 A large open space to be city park, at the center

表1　ICDCゲーラ2030 4つのプランニング要素（モジュール）
Tab.1 ICDC concept contains four main modules

統合的まちづくりコンセプトの4つの要素
concept contains four main modules:
　　なぜか？　総合的条件の設定
1　*Why?* defining overall conditions
　　どのように？　未来発展の戦略とガイドライン
2　*How?* strategies and guidelines of future development
　　何を？　プロジェクトにおける公・私共同活動
3　*What?* fields of public-private action with projects
　　何時か？　実施過程
4　*When?* implementation

図3　認定証を持つICDC市民
Fig.3 Ending scene of the meeting – together with "ICDC" citizen

2-6 富山市 -環境未来都市- コンパクトシティ・コンセプト
Toyama City Environment Future City & Compact City

中村 圭勇
NAKAMURA, Keiyu

■ 富山市のコンパクトシティ戦略

　富山市環境部の中村氏は、コンパクトなまちづくりを中心戦略に位置づけ、環境分野、高齢化対応分野も含めた先進事例を示して世界に発信している富山市の「環境未来都市」の施策を紹介した。（図１）

　富山市は、本州の日本海側のほぼ中央に位置し、北は富山湾、南は立山連峰を有する。製薬業を中心に地方都市としては恵まれた産業集積を誇り、人口42万人の中核都市として発展してきた。しかし、2045年までに約10万人減り、約32万人になると予測されている。高齢化率は27％で、2040年には40％を超える。全国平均を上回るスピードで高齢化が進むうえ、市域面積も 1241.85km² という広大な富山市は、もともとあった公共交通を活性化させ、その沿線に居住、商業、業務、文化等の都市機能を集積させ、「公共交通を軸とした拠点集中型のコンパクトなまちづくり」に取り組んでいる。3本柱は、(1) 公共交通の活性化、(2) 公共交通沿線居住の推進、(3) 中心市街地の活性化だ。

■ 全国初のLRT本格導入がもたらすもの

　中村氏は (1) 公共交通の活性化から解説。カギは、全国初のLRTの本格導入と上下分離方式による市内電車環状線化だった。

　富山市には富山ライトレール（2006年開業）、市内電車環状線化（2009年開業）がある。2015年3月には北陸新幹線が開通するので線路を高架化してこれらを接続させる予定もある。(図2、表1)

　富山ライトレールは、利用者減が続き、廃線が決定したJR富山港線を、公設民営の考え方を導入し、日本初の本格的LRTシステムとして蘇らせたもの。車に依存したライフスタイルを見直し、歩いて暮らせるまちを目指している。

　その成果はすでに現れている。開業前と比較して利用者数が平日で約2.1倍、休日で約3.5倍と大幅に増え（表1）、日中の高齢者の利用が増加。外出したくてもできなかった高齢者が、LRTによってライフスタイルが変わったのだ。

　一方、市内電車環状線化事業は、中心市街地の活性化とまちなかの回遊性強化を目的に、市内電車軌道を一部延伸。日本初の上下分離方式を導入したもの。「富山駅前と中心市街地が離れているため、環状線化でつないで経済の活性化を図った」と語る。環状線利用者は女性が約7割を占め、平日は女性の高齢者の利用が大きく増えた。環状線が日常の移動手段として定着している。

　「クルマを使わずに路面電車で来る人たちは、購入単価が高いうえ滞在時間が長く、アルコール類の消費も多いという結果が出ている。中心市街地の消費拡大に寄与している」とも胸を張る。

　LRTは、移動手段のみならず、まちを活性化していく1つのツールになることが明らかになった。

■ さまざまな施策で都市構造を転換

　(2) 公共交通沿線居住の推進については、公共交通沿線居住推進地区を設定し、補助制度を充実させることで効果が現れている。「公共交通を整備するだけでなく、公共交通の沿線への居住を推進するため、良質な住宅の建設事業者や住宅を建設・購入する市民に対してインセ

■ Toyama City and Machidukuri project background

Toyama city is located in the central part of Japan on the coast of the Sea of Japan, facing Toyama Bay to the north and surrounded by the 3,000 m high Tateyama Mountain Range to the south, which is blessed with an abundance of natural beauty. The city has developed into one of the major core cities along the Sea of Japan coastline because of the growing medical industry and the accumulation of other industries, and numbers 420,000 thousand people. Nevertheless, Toyama City's population is expected to decline 20% by 2040 compared to 2005. At the same time, its aging population is rapidly increasing at a rate exceeding the national average, with the number of elderly people over 65 predicted to be one in three in 2035.(Fig.1)

■ Challenges to do

In addition to the demographic changes, the city faces many challenges such as low population density caused by urban sprawl and "perforation" of the city center, the over-dependency on cars, and the decline of public transportation as well as the downtown area. Moreover, the city is concerned that these challenges will become more serious, and government expenditures will rise due to a population that is both decreasing and aging.

■ Strategies to realize the plan

The current mayor of Toyama City, Masashi Mori, who was first elected in 2002, realized the city's crisis situations as soon as he assumed office and resolved to follow a policy opposing drastic urban planning The result is "Compact City Planning," which the city is enforcing now. The three dominant concepts of Compact City Planning are as follows: (Fig.2, Tab.1)

1. Revitalization of Public Transportation
-The city introduced the first full-scale LRT system in Japan and an "Infrastructure-Operation separation" system for the loop-line tram that runs in the city center
2. Promotion of Living along Public Transportation Routes
-The city promoted and subsidized settlement in the areas along public transportation routes
3. Revitalization of Central Urban Area
-Concentrated investment by public sector and induced private investment, maintained land prices and the back-flow of taxes.

■ Leadership and global project approach

Under the mayor's strong leadership, the Toyama city was able to demonstrate clear images of "Compact City Planning." Those works were highly valued by the OECD, and the city is featured as one of the world's five most-advanced cities in the OECD report "Compact City Policies" in June 2011, drawing international attention.

But this is not the only reason the Urban Planning of Toyama garnered worldwide attention. In December 2012, the city was certified as a "Future City," one of eleven cities so designated in Japan by the national government. The "Future City" is an ambitious project that aims to create a successful and unique model of urban development in the field of environment and super-aging society through the cooperation of industry, government, academia, and the industry sector. It also targets national economic growth in

ンティブとして補助金を出している」と中村氏。2005年7月からは「まちなか居住推進事業」として共同住宅に100万円/戸、戸建住宅に50万円/戸、さらに2007年10月からは「公共交通沿線居住推進事業」として共同住宅に70万円/戸、戸建住宅に30万円/戸を助成している。こうした施策を通じて富山市のまちづくりに理解を示し賛同してもらい、公共交通沿線居住を進めていくことが狙いだ。中心市街地では2007年から人口が転入超過に転換した。公共交通沿線居住推進地区では増減はあるものの、転出超過が減る傾向にある。

（3）中心市街地の活性化は、公共による集中投資およびそれによる民間投資の誘発で地価の維持と税の還流を図っている。

「都市構造自体を転換する取り組みなので、行政の主導が欠かせない。公共投資が呼び水となり、市街地再開発事業など民間投資が活発化し、集合住宅への投資も増えている」と分析する。

また、2015年8月には、著名建築家・隈研吾氏を招いて建物をリノベーションした「ガラス美術館」をオープンする。北陸新幹線の開通を見据えた新しいまちの魅力づくりにも余念がない。

■首長の強力なリーダーシップ

富山市がこうした施策を積み重ねることができるのは、森雅志市長のリーダーシップが大きい。「LRTは計画されてもなかなか実現できないケースが多いなか、2002年に初当選した森市長は、就任と同時にまちづくりの抜本的な転換を決断した」と中村氏。それが「コンパクトなまちづくり」につながっている。

首長の強いリーダーシップのもと、富山市は極めて短期間に「コンパクトなまちづくり」の具体像を示すことに成功したわけだが、中心市街地や公共交通の沿線ばかりに目を向けているわけではない。「まずは中心市街地を活性化して地価の下落を抑え防ぎ、税収（固定資産税、都市計画税）を確保することで、市域全体に行政サービスを公平にお届けすると森市長は市民に丁寧に説明している」と言う。その結果、まちづくりへの市民参加も進んでいる。

こうした取り組みが評価され、2012年6月にはOECDが世界先進5都市の1つとして「コンパクトシティ政策報告書」に富山市を取り上げるなど、国際的な注目も高まっている。5都市のなかで、富山市だけが「人口減少下のモデル」である点もとして評価されたと聞く。

■「環境未来都市」構想でサステナブルなまちづくり

最後は中村氏が担う「環境未来都市」構想について。2011年12月、富山市は当時の国家戦略プロジェクト「環境未来都市」の選定11都市（被災地以外では5都市）に選ばれた。「環境未来都市」とは、環境と超高齢化分野で産学官民の連携によって世界に類のないまちづくりの成功事例を創出し、国内外に発信することで地域経済の活性化、さらに日本の経済成長にもつなげようというプロジェクト。富山市が選ばれたのは、コンパクトシティによる戦略的なまちづくりが地方都市のモデルになると評価されたからだ。

「チューリンゲン州でも都市部と農山村エリアをどう有機的にリンクさせていくか考えているそうだが、富山市も同じその考え思想を富山市環境未来都市計画に盛り込んでいる」と語る。（図3）

コンパクトなまちづくりの実現に加え、市民生活の質の向上と地域特性を活かした産業振興を図るため、①中心市街地・公共交通沿線での人口・諸機能の集積、②再生可能エネルギーと都市との交流・連携を軸とした田園・自然エリアの活性化について15のプロジェクトチームが取り組みを進めている。そのうち、小水力発電の電力を農村に利用する「再生可能エネルギーを活用した農業活性化プロジェクト」にインドネシア共和国バリ州タバナン県が興味を示し、2014年3月にプロジェクトの

addition to revitalizing the local economy by developing these success models at home and abroad.

In May 2012, Toyama instituted "Toyama Future City Planning" as the plan to implement Future City Initiatives. As the title "Establishment of Toyama-Type City Management Using Compact City Strategy—Toward a Sustainable City that Creates Added Value and Is Full of Social Capital" shows, this core strategy is also one of the "Compact City Strategies."

■ Next stage of sustainable Machidukuri in Toyama

In this stage of "Toyama Future City Planning," the city aims to achieve the compact city and to develop local industry in harmony with the area's distinctive characteristics in order to enhance the quality of citizens' lives. The plan also intends to create other values such as a concentration of population and many urban facilities in the downtown area or along public transportation lines; it also wants to stimulate the countryside by using sources of renewable energy and promoting urban–countryside interaction. The former shows the direction which the city aspires go in, that is, promoting local industries to achieve sustainable urban management in addition to accumulating urban functions and boosting the quality of citizens' lives. And the latter aims to integrate the urban management of public transportation, which combines urban functions with the countryside. To implement this city plan, fifteen programs have been prepared. (Fig.3)

By realizing these programs through the cooperation of industry, government, academia, and the private sector, we can assure that the "Toyama style urban management" and "The sustainable city that creates added value and is full of social capital" will happen just as the title says.

図2　富山市のLRTネットワークの形成
Fig.2　LRT network connecting to-vitalize points in the city

表1　富山ライトレールの整備効果（乗客数）
Tab.1　The LRT system is promoting citizen's mobility

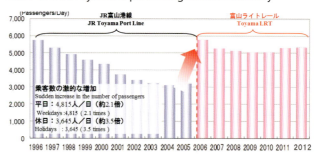

■ Number of passengers in use of the LRT 2006-2012

実施に関する協力協定を結び、プロジェクトに参画する民間企業が調査を始めたところだという。

コンパクトシティ化はエネルギー効率をアップすることにもなる。2014年9月、国際連合のSE4ALL（Sustainable Energy for ALL）が「エネルギー効率改善都市」（Global Energy Efficiency Accelerator Platform）として日本で唯一富山市を選定した。

「こうしたまちづくりを進め、持続可能な都市の条件といわれる環境価値と経済的価値、社会的価値の創造と向上を目指し、生活の質と環境が調和したサステナブルなまちづくりを進めていきたい」と締めくくった。

図3　富山市が目指す「お団子と串」の都市構造イメージ
Fig.3　The Image of urban structure connecting suburban areas by railways, bus service

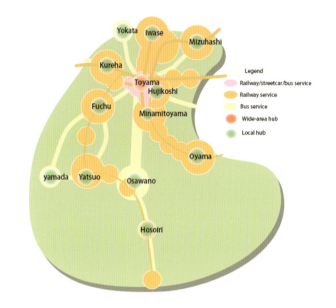

図1　富山市 - 雄大な立山連峰を背景に
Fig.1　Toyama City in front of Tateyama mountains
富山市提供著作権は富山市に帰属。　禁無断転載・複写

2-7 パッシブタウン黒部モデル
Passive Town Kurobe Model and Machidukuri

小玉 祐一郎
KODAMA, Yuichiro

■社宅跡地につくる「パッシブタウン黒部モデル」

建築や都市のパッシブデザイン、サステナブルデザインの研究開発と実践に従事している建築家の小玉氏は、富山県黒部市で現在進行中の住宅団地と商業施設からなる「パッシブタウン黒部モデル」について報告した。

黒部市は富山市の東側に位置する人口4万2000人ほどのまちで、トロッコ電車で有名な黒部峡谷や宇奈月温泉を擁する。また、ファスニング事業、建材事業、工機の3部門から成り、世界71カ国・地域に108社の拠点をもつグローバル企業・YKKグループの本拠地でもある。

パッシブタウン黒部モデルの計画地は、JR黒部駅と間もなく開業する北陸新幹線・黒部宇奈月温泉駅の間にある。そもそもYKKグループの社宅だった場所だ。(図6)

「この社宅の建て替えの話が出たとき『たんなる建て替えではなく、まちに開き、地域の活性化につながるようなプロジェクトにしたい』というYKK側の強い意向があった」と小玉氏は明かす。

約4haの社宅跡地に順次建物を建てて、多くの人に使ってもらい、そして黒部市の活性化にも貢献するという狙いで出発した。10年ぐらいかけて250戸ほど建てる計画で、今は第一期の工事中だ。

■太陽や風など「ローエネルギー」を活かす

パッシブタウン黒部モデルは、「21世紀の持続可能な社会にふさわしいローエネルギーのまちと住まい」を目指す。

「太陽・風・緑・水といった黒部の自然ポテンシャルを活かすことが大きな比重を占める」と言う。

黒部市は黒部川の扇状地だが、やや特異な気候をもつ。日本海側なので冬の日射は多くないが、夏は暑い。冬は南から冷たい季節風が吹き、逆に夏は北から心地よい季節風「あいの風」が吹く。扇状地なので地下水が豊富に湧き出る。また、森林資源も豊かなので、バイオマスエネルギーのポテンシャルも高い。そのような気候風土を活かし、なるべく在来エネルギーに依存せず、快適な暮らしを実現することがパッシブタウン黒部モデルの基本的コンセプトだ。

低密度ではあるが、豊かにある「ローエネルギー」を活用すれば、ガスや電力などの高密度・高品質な「ハイエネルギー」を、必要とする用途に適切に有効に活用するという持続可能な構想が生まれる。地域との共生にも貢献する。

「なぜパッシブデザインなのか。1つはハイエネルギー消費の削減だ。これはグローバルな地球環境問題への必須の対応である。もう1つは、地方都市のこれからのライフスタイルとして、自然との共生や交感によってローカルに解決していくことが、地域のコミュニティづくりにも大事だからだ」と主張する。

スイッチのオンオフで得られる手軽な快適さよりも、地域の資産をうまく活かして自らつくる快適さを実現する——これが狙いである。(表2)

アクティブな生活を送るにはパッシブデザインが適していると考えている。風土を活かしたパッシブデザインで自然の快適さ、暖かさ、涼しさを取り入れて持続可能な社会を築く。アクティブな住み手が参加することで地域のコミュニティも開かれていく。「こうしたことに貢献できるのがパッシブデザインだ」と語った。

■ Passive Town Model will generate sustainable Kurobe city

At a corner of Kurobe city, an eco-housing complex is now under construction. This housing is designed as a passive eco-system, consisting of about 40 apartments. In the coming years the project will be expanded to 260 apartments with a community center.

Kurobe is a small city with a population of about 42,000 on the coast of the Sea of Japan, nestled at the foot of Mt. Tateyama. The housing is being built for the employees of YKK company, which is a world-renowned manufacturer of aluminum products. Since its founding, YKK adopted the policy "Think globally, act locally."

YKK is building this eco-housing as a model for living with the local environment. The project expects the private and public sectors to work closely together.

■ Eco-building design based on local climate conditions

The climate in the Kurobe area is unique in terms of sunshine, wind, the green landscape, and water, which is available from underground and hot springs (Fig. 6).

■ Energy use planning based on low-energy and high-energy

The low, meaning low density, energy is used for purposes such as heating water, which does require much energy. Residents themselves are to manage energy use for their daily living, their "active life" (Tab. 1).

■ The housing will generate interaction of dwellers and citizen

The apartments, balconies, "community decks," and open spaces provide opportunities for social life. (Fig. 1, 2, 3, Tab.2)

表1 エネルギー削減のシナリオ
Tab.1 Scenario of saving energy living in the housing

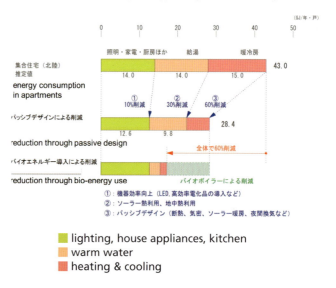

■エネルギーを使い分ける

では、ハイエネルギーよりも低密度でエネルギーとしての質では劣る太陽や風、地中熱など地域にあるローエネルギーをどう使えばよいのか。小玉氏は「使い分け」を提案する。

「ハイエネルギーしか適用できないところはハイエネルギーで、ローエネルギーで賄えるところはローエネルギーで活用する。特に建物の場合は、品質の悪いエネルギーでも使える用途がある」と言う。

ここで北陸の集合住宅のエネルギー使用比率を推計した「エネルギー削減のシナリオ」を投影した（表1）。

「家庭では年間およそ43.0ギガジュールのエネルギーが使われている。照明・家電・厨房、給湯、暖冷房がだいたい1/3ずつだが、北陸は暖房に使う比率が少し高い。といっても暖房の充実しているドイツの1/3だ」と説明する。

今回のパッシブタウン黒部モデルは、暖房を充実させたうえで、それに費やすエネルギーを現在の60％、給湯は30％、照明・家電・厨房は10％それぞれ削減するという目標を立てている。さらに将来的には水やバイオマスなども使うことで、全体で現状の60％を削減しようと考えている。

「大幅なエネルギー削減と自然に親しむ心地よさを両立するまちづくりを考えていきたい」と述べた。

■快適でまちに開かれた住宅棟

では、「パッシブタウン黒部モデル」の具体的な計画はどのようなものなのか。投影された第一期開発計画の配置図では、商業棟と住棟（36戸）から構成される。（図1）

住棟は、躯体の中心部はしっかり断熱を施したRC構造で、南北の外側は鉄骨造にする計画だ（図2）。北側は夕涼みなどで人々が集うコミュニティデッキを設ける。南側は住戸ごとにテラスを設けてアウトドアリビングスペースとする。

「黒部市の暑い夏を快適に過ごすには、住戸に風を取り入れる装置と、半公共のテラスを設けている」と説明する。

前述したように、冬場は南から冷たい風が吹くため、住棟の南側は風よけのガラススクリーンを配置しつつ、サンルームのように太陽光を楽しみ、また室内に取り入れる（図3）。また、テラスには可動式のルーバーを設置し、「晴れの日はルーバーを開放し、雨が降ったら閉じる。伝統的な土間のような使い方ができるようにも考えている」と話す。

駐車場は、1戸につき1台の割合で地下に設けている。

「駐車場を地上に設けて、重要な外部空間を損ねることはしたくなかった。しかし現在のこの地域では車の利用は欠かせない。未来のライフスタイルを考えて地下駐車場にした」と説明する。車の雪対策も容易になる。将来はカーシェアリングなどへの移行も考えられるので、そのときの対応にも考慮している。

これによって、2つの住居棟の間にある屋外スペースは緑の共有空間になる。将来的には、敷地の一部を一般市民に開放する公園にするという計画も進めている。北側に流れる川を活かして親水環境をつくり、南の道路側のショッピングモール→商業棟とつないで、まちに開かれたコミュニティになるような設計も施している。（図4）

「北側の川から地中熱や水を活かしてエネルギーを生み出すと同時に、親水空間も室内に取り込んでいきたい」と述べ、パッシブタウン黒部モデルの紹介を終えた。

表2 アクティブな生活のためのパッシブデザイン
Tab. 2. Passive design environment stimulates active life

PASSIVE DESIGN FOR ACTIVE LIFE

アクティブな生活のためのパッシブデザイン
地域の風土の特性をいかす
自然の快適さ・暖かさ・涼しさ ー パッシブデザイン
‖
持続可能な社会＝省エネルギー
‖
住み手参加、ライフスタイルの尊重 ー アクティブライフ
地域のコミュニティに開く

use of local climate potenntials
natural comfort, warmness, coolness
sustainable community =energy saving
dwellers participation, new life style
Open to the surrounding communities

図1 第一期開発計画 配置図
Fig.1 Site Plan and floor plans, the 1st development phase

図2 住棟の構造（黄色がRC造・外断熱、青色が鉄骨造）
Fig.2 The building is consisted of two parts

図3 太陽光と北風の取り入れ方法
Fig.3 Building design to get sunshine and wind into rooms

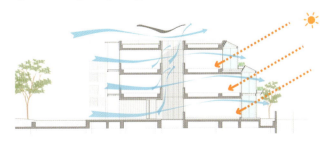

図 4　北住棟南側のテラス（左）と南住棟北側のコミュニティデッキ（右）に囲まれた中庭空間。
　　　地下に駐車場が設置されている。奥に、道路に面した商業施設棟が見える。
Fig.4　The space between balcony and "community deck"

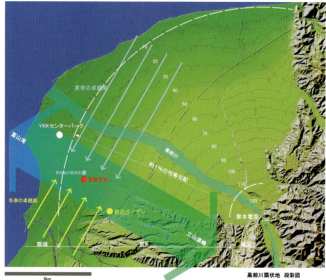

図 5　黒部川扇状地 - 立山の裾に広がる扇状地は特有の気象条件である
Fig.5　Alluvial fan of Kurobe River-Landscape and climate

図 6　立山連峰の北側に広がる黒部のまち。広大な扇状地の中央に黒部川が流れる。（提供: YKK）
Fig.6　Tateyama mountain and the city of Kurobe

2-8 討議とまとめ - 日本の都市成長、コンパクトシティのプランニング、合意形成の実態

Discussion and Conclusion – Urbanization, compact city concept and consensus building

澤田 /SAWADA、大村 /OMURA、シュトレープ /STRAEB、猪股篤雄 /INOMATA、門脇耕三 /KADOWAKI、永松栄 /NAGAMATSU

■ 重厚長大産業によって発展した日本の社会・都市

澤田 プレゼンテーション全体を振返って、今後のまちづくりの方向性を考えたい。

門脇 日本でのサステナブルなまちづくりは、公的コントロールが難しく自治体財政も "壁" になっている。だからといって放任するわけにもいかない、というのが現状だ。富山は人口配置を戦略的に進めていて参考になる。現在ストックが飽和したので、従来のような開発誘導型ではなく、開発の概念そのものを変える試みをしている。「箱の産業から場の産業へ」という方式は建築産業体制や都市計画手法が従来とは違って来ることを予感させる。これが普及すれば人口再配置の大きなツールになるだろう。

猪俣 ドイツとの比較で「サステナブルなまちづくり」を考えるのであれば、"一極集中の日本" 対 "そうでないドイツ" とか、フランクフルトやベルリンは大都市だが、東京・大阪に比べればずっと小さいとか、教育制度の違いなどを、もっと良く見ないといけない。

日本は、一極集中の産業・経済活動によって右肩上がりの時代に世界トップにまでなった。エネルギーもないのに人的資源だけで達成したのだ。それが1985年のプラザ合意以降、あっという間に下降し始めた。こんなことは東京にいると気づかないし、地方でも感じない。日本におけるサステナブルまちづくりは、地方対東京の関係で見るだけではなく、日本全体を見て考えなければならない。それも2030年までにキャッチアップが必要だと思う。ドイツ並みにならないといけない。

■ コンパクトシティのプランニング

澤田 "サステナブル社会のまちづくり" を比較しようとすると、視点は2つあるのだと思う。"コンパクトシティをどう定義するか?" と "合意形成はまちづくりのプランニング・プロセスの中でどの位置づけになっているのか?" という点だ。

日本の地方都市の中では富山市は、サステナブルまちづくりのガバナンスが特別強い事例だ。これを考慮するとどうなのか?

シュトレープ まちづくりプロジェクトでは、人口の増減や流動という変化が大きな問題だ。それに対して自治体がどのような目標を定めて、投資の誘導を計画する。そうしたプランニングを始める際の合意形成は、どのような関係者をどのように巻込むかという活動だから、一番重要な活動だ。富山の事例ではそうした方向性を明確にして、いろんな制約のある中で実現した事例であり、自治体に課せられた役割をはっきり示している。

中村（富山市） 富山のコンパクトシティは、人口減、高齢化の中で広大な市域を持つので従来型の郊外開発を続けると、都市の維持のコストバランスが崩れる。降雪地なので除雪などを含むインフラコストも膨らむ。人口減少に合わせて都市の規模を適正化しなければならない。

高齢化を合わせて考えると、広い市域を行き来し易くするための公共交通が必要、また、そのLRTの沿線に住んでいただくという計画だ。つまり都市の特性と地域のもつ資源を合わせた。

市主導の戦略をコンパクトシティと呼んで計画し実施した。

■ The Social Development of Japan and its Industry

SAWADA Looking back at the presentation, I would like to review the future of Machidukuri.

KADOWAKI What is proving to be difficult for the development of sustainable Machidukuri in Japan is the "wall" created by central government's control over local government through financing. Something has to be done to change the current situation, and Toyama is a good example of where the strategically positioning of the population can be observed.

Due to the present stock saturation, we cannot continue the current induction development method; what is perhaps required is a change in the approach to the development concept itself rather than its methods.

INOMATA When comparing "Sustainable Machidukuri" of Germany with that of Japan, we can see differences between the two, such as the "densely concentrated nucleus of Japan" vs. the "decentralized cities of Germany."

Other aspects such as the differences in the educational systems should also be considered and compared. Japan has a geographical landscape that extends from north to south, and in order to overcome the geographical limitations, the Japanese government centralized its economy.

The centralized policies turned Japan into one of the top economic powerhouses during its economic growth period.

It's hard to perceive these matters when living in Tokyo or the countryside, but they can be strongly felt in the prefecture of Kanagawa. Kanagawa is following a different path of de-industrialization, and I want the people in Kanagawa to understand the need for searching for a sustainable option to the de-industrialization of their city. The sustainable Machidukuri in Japan is a matter that needs to be seen as more than the relationship between city and countryside; it is a matter that must take into account the whole country and one Japan must try to resolve by 2030.

■ The Planning of Compact Cities

SAWADA The definition of a "compact city" and where the matter of "consensus" fits into the planning process are perspectives that have to be considered when evaluating the "Machidukuri of a sustainable society." The city of Toyama is a rural city with strong governance regarding the matter of sustainable Machidukuri that should be considered for debate.

STRAEB In a Machidukuri project, the fluctuation of the population might cause major problems, and the local government must clarify its aims to tackle these issues in order to attract investors. One of the most important actions is arriving at a consensus during the planning process and the method in which concerned members are brought into the process.

In the case of Toyama, the project's goals were made clear, and even though the project had many constraints, the local government was capable of planning the project by clarifying its role.

NAKAMURA（Toyama City） Toyama has vast city limits, and with the issue of a decreasing and rapidly ageing population, conventional suburban developments have become an inappropriate

ミラー（ゲーラ市）　LRTによる短距離移動の快適さはコンパクトシティの特性の一つで面白いと思う。短時間で移動できれば、ほかのことをする時間も生まれる。自治体としてはインフラ投資を減らすにはできるだけ中心部に住んでもらえるようにしたい。特に低成長時代の都市では短距離で動けるのがコンパクトシティの要件だと、自分は考えている。

■プランニングの進め方（合意形成）の実態

澤田　ドイツでは、今まで市民はタウンプランニングにあまり関心がなかった。大抵は自分の家の前のことだけに限られていた。最近になって状況が変わり、ゲーラ市では150名の市民などが2年の間ボランティアで参加している。自治体が飲み物の費用程度は負担することで、市民の参画を促している。会議を開催する自治体側としては、コストはかけないが、時間はかける必要があると考えている。

大村　日本のまちづくりでは、大学の研究室がワークショップを開き地域の人々とアイデアを出し合ってプランニングを進めることも多い。その際あまりコストをかけないでやっている。

永松　最初の1年は復興計画のフレームづくりなのでいわゆる素人は議論に参加しない。今は落ち着いて自分たちのまちを計画する段階に入り、外からNPOやコンサルタントを呼んでワークショップをやっている。自治体としても意味のある会合の費用は負担するように変わってきている。

猪俣　まちづくりの始めに市民との対話が必要だ。日本では昔から町内会というものがあり、それを接触点として、自分のビジネスもやっている。先日もシンガポールの公社の人が勉強に来た。類似の組織はドイツにはないのだろうか？

シュトレープ　そういう仕組みはドイツにはない。企業が住民の意見を聞くことはいろいろな形で行われている。居住者組合がインフォーマルに、動くケースもあるようだ。しかし、まちづくりで大事なのは、自治体、投資家、住民組合が積極的に動いて、各立場の利害をはっきりさせることだと思う。それがプランニングにとって一番大事だと自分は考えている。この点で、日本はどうなのか？

大村　団地再生については、日本の団地ではおなじ時に似た階層の住民が集まり、当時は一緒に町をつくるという意識があったが、今では世代も代わり、考え方もさまざまになってしまった。団地再生プロジェクトの合意形成では、住民意識そのものがよみがえるのを待つ必要がある。

前川（聴衆）　ドイツではどのようなモチベーションが設定されるのか？サステナブルの"3つの柱"を同時に詰めようとしているのか？ゲーラ・プロジェクトではどうなのか？

ミラー（ゲーラ市）　テーマ別に編成されたグループごとの意見のまとまり方は、自分はいつも見て回る。市民はインターネット（Uチューブ）も使って意見の集約と評価に取組む。それでも"コインの裏と表"の喩ではないが、表面で"やるべきこと"を見て、裏面では"どう実行するか"を見て議論することになる。自分はそうした状況を"他のまちと比べて良くやっている"ときちんと評価することにしている。

澤田　日本では"大都市の中心部"、"大都市の郊外"そして"地方"という場所ごとにコミュニティのあり方は違うように思う。サステナブルなまちづくりでのモチベーションの付け方はそれらお場所に応じて多様なのだと思う。

planning option. When considering the issue of a rapidly ageing population living in such a large city area, we can understand that public transportation systems are a basic necessity. And in order to optimize these public transports, residential areas must be located along them.

MILLER（Gera City） I think that the short-distance LRT is an interesting characteristic of a compact city that reduces travel time. Also, the centralization of residential areas makes for lower infrastructure costs and suggests that short-distance transportation is indispensable for a compact city during periods of low economic growth.

■ The realities of consensus in the planning process

STREAEB Until recently, Germans have shown little interest in town planning, but the situation has changed recently. In the city of Gera, 150 residents have come together and volunteered during these past 2 years. From an organizer's perspective, it is a process that requires time rather than financial aid.

NAGAMATSU The framework was created by a professional party during the first year of the reconstruction project. The project has since then settled and has entered a phase where NPOs and consultants are being introduced to the project to take part in the city planning.

INOMATA Japan traditionally has had a strong local community. It is a positive sign that people are volunteering. Recently, officials from Singapore visited us to learn about our traditional system.

STREAEB In Germany we do not have a system like the one in Japan, but there have been many cases where companies have made surveys. The important factor in a Machidukuri project is the clarification of interests between local governments, investors, and residential associations.

OMURA Similar incentive was organized by Japanese Housing Corporation in an effort to involve the local residents in the Machidukuri process. But now with the passing of generations, the mind-set within the organization has diversified. I believe that in order to summarize the opinions of the residents, the housing estate refurbishment project must wait until a people's consensus is reached within the community.

Audience Would like to ask what the general motivation of the citizens is for the Machidukuri process in Germany and also how close the Gera project comes to achieving the objectives of the 3 pillars of sustainability.

MILLER（Gera City） Tools such as the Internet are used to gather opinions and evaluations from citizens. This information is then organized into specific topic categories.

SAWADA In Japan there is a difference between how communities function in "large city centers", "suburbs," and "regional areas," and I believe that the motivation for the Machidukuri is different in each of these sites.

第3章 神奈川・横浜で取組む生活・産業・住環境の再構築
Chapter3　Development of Life, Industry and Housing in Kanagawa Yokohama

首都圏と神奈川
Tokyo Metropolis and Kanagawa Prefecture

横浜会議　Yokohama Conferense

猪股篤雄（文責：前川）
Mr. INOMATA
戦後70年の神奈川のまちづくりの変遷と
これからの取組み
Machidukuri Development in Kanagawa
Region and Future Strategy

箟建夫（文責：前川）
Mr. SHITOMI
汐見台団地の計画と成長の経緯、
これからの課題
Shiomidai Housing Estate Project –
Overview of its History and the next Programs

鈴木浩伸（文責：前川）
Prof. SUZUKI
神奈川における都市計画の発展と現状
Town Planning and its Issues of
Kanagawa Area

討議とまとめ -
現場を見てポジティブに考える、
コンパクトシティの意味、国際交流のあり方
Discussion and Conclusion –
Positive thinking on site, Compact City
prospect, International exchange

横浜市ー市街化エリアと交通網
Yokohama City – Urbanized areas and transit networks

3-1 戦後70年の神奈川のまちづくりの変遷とこれからの取組み
Machidukuri Development in Kanagawa Region and Future Strategy

猪股 篤雄
INOMATA, Atsuo

■ 限界を迎えた日本の社会システム

2日目の午後は横浜市中区の神奈川県産業振興センターに会場を移し、神奈川県住宅供給公社（以下、公社）理事長の猪股氏が登壇。神奈川県の戦後70年間を振り返り、持続可能なまちづくりについて講演した。

公社は県下に約1万4000戸の住宅を抱えているが、県西部の住宅になかなか人を呼び込めないという問題も抱えている。しかし、猪股氏は「それは住宅だけ考えていけばよいというものではない」と主張する。「戦後長く続いた右肩上がりの経済情勢ならば住宅のことだけを考えていればよいが、今のような低成長時代では住宅の長寿命化および建て替えだけではニーズに対応できない」と語る。そこで、まずはまちの成り立ちに深くかかわる産業に着目し、産業と住宅の関係性について述べることが主旨だった。

今から15年先の2030年、日本は産業の空洞化、消費の減衰、人口減少、少子高齢化、労働人口減少などさまざまな問題が表面化する。2030年には家族形態が変化し、一人暮らしは人口の40％近くに達し、そのうち40％は65才以上の高齢者となる。1万カ所の集落が高齢化率50％以上となり、消滅するともいわれており、空き家も増加の一途をたどる。

戦後70年は大量生産と大量消費の時代だったが、多くの環境問題や資源問題を引き起こした。生産と消費のバランスがとれた持続可能な社会への転換が求められている。「実は、名目のGDPは2010年から下がっていて、生産人口も2010年から減少していることを認識すべきだ」と指摘。大量消費や高度経済成長期を支えた団塊世代が引退し、その子どもたち（団塊ジュニア）もすでに40歳前後となり、旺盛な消費が期待できる時期は終わりに近づいている。「今の日本の社会システムの限界がすでにきているのかもしれない」と危機感を露わにした。

■ 30年前からはじまっていた産業の空洞化

日本の国土は南北に長く、およそ3000kmにわたって延びている。これはドイツからギリシャよりも長い距離だ。日本には地域によってさまざまな状況がある。

「神奈川県は、江戸時代から一極集中化を続ける東京の隣にある。この神奈川固有の産業と住宅の関係を見極めて15年先を考えることが重要だ」と話す。

神奈川は東京のベッドタウンと思いがちだが、実は歴史的な生産拠点としての製造工場を有する。原料を輸入して生産し、製品として輸出する日本のビジネスモデルの一翼を担っていた。午前中に訪れた汐見台団地は、神奈川臨海部の埋立地に進出した企業の従業員の受け皿として建設されたもの。つまり金融と本社機能が一極集中した東京とは異なり、神奈川には歴史的な生産拠点として製造工場が集中し、それに伴う住宅開発が行われてきたのだ。

神奈川は「戦前からの工業地帯」だった。（図1）戦後は京浜工業地帯と軍需産業地帯を背景に、朝鮮特需により飛躍的な復興を遂げる。川崎と横浜は港湾機能をもっていた。軍港があった横須賀はアメリカ海軍に引き継がれ、それに伴って軍需産業地帯が発展した。その受け皿となったのは、横須賀の海軍の大砲のための火薬を明治時代から製造していた厚木と平塚。そのノウハウは高い化学薬品製造に引き継がれる。

■ Japanese social systems are reaching the limits.

A brief overview of the development of life, industry, and housing in Kanagawa Prefecture following WWII shows that the social system of Japan is now approaching "the limit." Kanagawa has a population of 9.1 mill., including the port city of Yokohama with 3.7mill. people and Kawasaki City (1.5 mill.), an industrial town. They form a major part of Japan's capital region. KPHSC owns 14,000 housing units, and is hurt by the problem of vacant apartments in the western area of Kanagawa. This downturn in the Corp.'s business was influenced by the industrial development in the capital region. During the period of economic growth, the Corp. should have developed housing estates, buying land, building housing, and operating it. With the onset of low economic growth, however, it was no longer easy to satisfy the users' needs by extending housing services.

The Corp. should now focus on economic activity that is beneficial even in a no-growth economy. This means town planning is essential for the business. It is said that 15 years from now, in 2030, Japan will be facing various critical issues such as the "hollowed-out industry syndrome," declining consumption, low birthrate coupled with an aging population, and a reduced labor force.

Also in 2030 family structure will have changed. About 40 percent of the population will be living alone. Many people will be over 65 years old. 10,000 rural communities will have disappeared by more than 50 percent because of an aging population. The number of vacant houses will rise.

The continuous mass production and mass consumption during the 1970s caused environmental problems and depleted natural resources. Now is the time for us to shift to sustainable society. Being aware of this means understanding that nominal GDP and the workforce have been in decline since 2010. Baby boomers who supported mass consumption and high economic growth are now retiring, and the next generation (children of older boomers) are around 40 years old.

■ Hollowing out of industries

Kanagawa's position in the capital territory is unique. In Tokyo, where finance centers and corporation headquarters are concentrated, Kanagawa has been a production base with a number of manufacturers. The housing units were developed to accommodate industry workers. As seen in Fig.1, Kanagawa was an industrial region prior to WWII. After the reconstruction following WWII, the area revived again because of special demands arising from the Korean War. The Keihin Industrial Belt area and the military industries were developed during that time, and Kawasaki and Yokohama were made into ports. The military port in Yokosuka in the south was taken over by the US Navy. The demands from the US Navy stimulated Yokosuka's military industry. Atsugi and Hiratsuka in the south on the Pacific coast once served as ammunition bases to meet navy demand beginning in the Meiji era. Their technical know-how, which had high standards, was inherited after WWII by the local chemical industries (Fig.2) . During rapid economic growth, the Keihin industrial area expanded, the chemical industry in Hiratsuka and Odawara became active, and the car industry became accumulated in Atsugi. The concentration of the population in Tokyo sharply raised land prices, and a number of housings estates were developed in the suburban areas in the capital region, most by public agencies. Then the bubble burst, bringing the period of

高度経済成長期は、京浜工業地帯の拡大と延伸があり、平塚と小田原では化学薬品産業が盛んになり、厚木には自動車産業が集積した。東京への人口集中と地価高騰によって遠隔地住宅（団地）が増えていく。そしてバブル経済が崩壊し停滞する。（図2）

神奈川の産業を見ると、実は1985年の「プラザ合意」による円高で空洞化がすでに始まっていた。（図3）

「驚くかもしれないが、産業の空洞化は30年前から起きていた。昨日の講演でも紹介されたように、旧東ドイツの都市はたった数年で、しかも急激に空洞化したからわかりやすかったかもしれない。しかし神奈川の空洞化は30年かかっているため見えにくい。気づいたら大変なことになっていたというのが実態だ」と語る。

■ 団地と世界政治はつながっている

年代順に見ていくと、重要だったのは1ドル240円が1年後に1ドル155円になったプラザ合意。これによって産業の海外移転、県外移転が始まり空洞化が起きた。その後のバブル経済はソ連の崩壊とともにはじけたが、止めを刺したのは1993年のBIS規制適用（バーゼル合意）。これによる銀行の融資規制が日本経済を一気に下降に転じさせたと分析する。（図5）

まちづくりを考えるシンポジウムでこうした話をするのはなぜなのか。「経済行為の上に政治的な動きがあるからだ」と猪股氏は言う。

「アメリカはソ連が崩壊する10年前からそうなることがわかっていた。アメリカにとって次の敵は日本だ。それでプラザ合意に踏み切った。実は、プラザ合意ではアジア通貨をそのままにして円高を実施したが、本来はドルを切り下げるべきだった。ところが円高にしたため、その後に韓国と中国が輸出で躍進することになった」と解説する。

その後、1991年にソ連が崩壊し、アメリカが世界を制すると思ったが、2001年にアメリカ同時多発テロ事件が起き、イスラム世界の拡張や中国の海洋進出などもあって、多様化の時代に突入した。

「神奈川の団地では建物の老朽化が進み、住民が減って高齢化が進んでいるが、これはたんに地域の問題だけではなく、世界の政治や経済の動きと連動している。世界から地域社会まで、実はずっとつながっているという視点をもった方がよい」と述べた。

■ GDDP（団地内総生産）で「見える化」を

神奈川では京浜工業地帯の衰退と化学薬品産業の移転が進み、自動車や家電などの製造拠点も県外や海外に移転した。こうした製造工場移転には労働人口の移動も伴う。

「厚木辺りでも今、建物がどんどん建っているように見えるが、あれは物流倉庫が増えているだけだ」と言う。

高度経済成長期に日本各地から神奈川に移住した世代の老齢化も進んでいる。地価の下落によって東京圏（東京・川崎・横浜の一部）への都心回帰の動きも見られる。

産業の空洞化、住民の少子高齢化、さらに住宅（郊外団地）の老朽化と、神奈川県の現状は厳しい。神奈川における団地住民は、65歳以上の高齢者率が県の平均値より10％以上も高く、超高齢化が進んでいる。産業の進出とともに開発された大規模な住宅は、産業が縮小するなかで取り残されている。住宅をどのように再構築すればよいのか。公社は何を目指すのか。

それに対して「資源をもたない日本人が人的資源を大切にして経済大国に登りつめたように、団地も人的資源を活用したい」と語る。

住民による自治会活動は今も機能している。その人的資源を活用し、団地内の経済活動を住民とともに高めたい。そう考えている。

ここで猪股氏は「GDDP」というオリジナルの指標を提示した。

「GDP（Gross Domestic Product、国内総生産）指標を団地にあ

rapid economic growth to a sudden halt. The economy continues to stagnate. However, following the signing of the Plaza Accord on September 22, 1985, the Japanese yen soared, which triggered the "hollowed-out industry" syndrome (Fig. 3).

Mr. Inomata noted that the hollowed-out syndrome had already begun 30 years ago. This social change in Japan was very similar to that in former East Germany, but the Japanese people did not take the problem seriously since it has proceeded so slowly over the course of about 30 years.

■ Housing business closely linked to world business.

Because of this, the Japanese companies started to relocate to areas outside Kanagawa or overseas. The burst of the bubble economy occurred again when the Soviet Union collapsed in 1991 and the BIS regulations in 1993 (Basel Agreement) delivered the death blow. Bank lending regulations caused a dramatic downturn in the Japanese economy (see Fig.5). Mr. Inomata thinks that this kind of global political change should be included in the symposium discussion since the title, after all, contains the word "sustainability," which refers to the long-lasting, stable operation and management of housing. He also emphasized that "local society and world are linked" today; "there's no getting around it."

■ Ways to encourage initiatives toward housing development

In Kanagawa, the chemical, automotive, and consumer electronic industries have moved away to other areas. The workers in these industries have also relocated. When housing was constructed, people moved to Kanagawa from other regions in Japan. This happened during the period of rapid economic growth, and these people are now aging. Due to declining land prices after the bubble, people moved to other parts of the capital territory—sometimes also to large housing estates in the suburban areas. (Tokyo, Kawasaki, and a part of Yokohama).

The current situation of KPHSC housing is critical because of the hollowing-out of industries, low birth rate and aging population, and housing deterioration. In the housing estates, the number of senior citizens over 65 is 10 percent higher than the prefectural average. Large housing complexes that were built during the period of economic growth 30 years ago have been abandoned in the shrinking economy. We should challenge this serious problem as well. As Mr. Inomata stated, "Although Japan is a poorly endowed nation, it has become one of the strongest economic powers by utilizing its human resources. We should encourage communities to do the same. Community activities at housing complexes are still well facilitated." By encouraging participation of "human resources (residents)", we could revitalize "economic activities" in the housing estates. He introduced a keyword or index "GDDP," which means "Gross Domestic Product in Danchi (housing complex)," to illustrate the "economic activities." He introduced several residents' programs whose purpose is to promote beneficial activities in the housing complexes. They keep production and consumption closer in Danchi, which might increase the GDDP even if only by a small amount. Also the programs use the results and achievements for recruiting new residents. By promoting closer contacts between senior and new young residents, they aim to make living at Danchi more attractive. The programs will show forms of sustainable community in and outside Danchi.

An example at Wakabadai Danchi is a summer festival that attracts about 50,000 visitors. If each visitor spent one coin (JPY500), the business from this event would show what the GDDP of Wakabadai is. (The festival could include a fireworks display, costing JPY8 million, which would surely lead to an increase in consumption in Danchi.) Another example: a basic energy infrastructure is already in place in Danchi, as are schools and hospitals. The facilities are set up like a "compact city." What Danchi still lacks is a profit-making entity, such as a manufacturing plant. The Corp. established an office

てはめて、GDDP（Gross Danchi Domestic Product、団地内総生産）として団地内の総生産を見える化する狙いがある」と解説。

「団地内の生産と消費」のバランスをとりながら総生産を高めて、問題点を住民とともに明確化し、さらに新たな住民を呼び込む。こうした地域における住居バランスをとりながら、持続可能なかたちにもっていくことを考えているのだ。

「私たちが管理・運営している若葉台団地には、5万人が集まるといわれている夏祭りがある。5万人全員がワンコイン（500円）使うとどれくらい消費が増えるだろうか。そして、会場で花火を上げるのに800万円費やしている。それは団地内の消費が増えていることを意味する」とGDDPの考え方をわかりやすく解説した。

団地には、基本的なエネルギーインフラがしっかり整備されている。学校があり、病院もある。コンパクトシティと呼んでもよいかもしれない。唯一整っていないのは製造工場のような収益を上げる機能だけ。そこで公社は新たな収益源としてオフィスをつくり、生産性を上げるにはどうしたらいいかを模索している。

「私たちが保有する県内の団地を有機的につなぎ、相互補完しながらグリッド化し、神奈川全域としてのエネルギー効率、消費効率を上げてゆく。団地全体としてエネルギーの再生・再構築を行ない、消費をどこまで高められるかに取り組んでいる」と語る。こう考える公社の「サステナブルプラン」に則って、10メガワットのメガソーラーも2015年4月から売電をスタートする。（図4）

■ 住民を元気にする独自策

一方、団地の活性化には、高齢者の健康、そして若年・子育て世代の流入が必須だ。

「団地は、できたときは若い年代が多いけれど、30年経過すると子どもたちは成長して出ていってしまう。それは当然のことだ。けれども常に右肩上がりの時代だったため、考える必要はなかった。しかし今は右肩下がりの時代なので、しっかり考えなくてはならない」と語った。

高齢化が進むと消費は減退する。「たとえば、4人家族だとリンゴを4個食べるけれど、老夫婦だけになるとリンゴは1個でよくなってしまう。そうして消費が減退し、空き店舗が増えてしまう」と言う。どうしたらよいのか。

公社は、団地再生事業の一環として、次世代を見据えた郊外型団地再生の可能性を高めるため、いくつもの取り組みを進めている。

横浜市旭区の若葉台団地では、中心商店街の活性化に向けて「職」と「食」に焦点を当てた「コミュニティ・オフィス＆ダイニング 春（Haru）」を2014年4月にオープンした。また、2014年5月には、0〜3歳の未就学児と保護者の親子が気軽に集まり自由に過ごせるひろばとして「わかば親と子のひろば そらまめ」も開所。いずれも若葉台団地商店街・ショッピングタウンわかば内にある。

これによって、シニア世代の生きがいづくりと団地内の若年・子育て世代の流入を目指し、「持続循環型のコミュニティ」を実現しようとしている。

「状況は厳しいけれど、こうした取り組みを広げて、公社の団地から『ボトムアップ型』の持続可能なまちづくりを目指していきたい」と述べた。

facility to find a way to increase productivity, namely, by providing work to the residents. The Corp. will soon connect compact cities (KPHSC-housing estates) together and increase consumption efficiency. The Corp. decided to create a power generation business and supply it to the housing estates. "Danchi-Grid" is the Corp.'s plan for its housing refurbishment project. The grid connects estates, first in order to supply electric power to each one, and also to pass on information about new businesses in and among the housing estates. A 10-mega watt solar electric generating plant will be in operation in April of 2015, with power being supplied through the grid (Fig.4).

■ Unique ways to attract the residents' participation

In most of the Corp.'s housing complexes, there is an urgent need to have residents from the younger generation and to provide quality care for seniors. "When housing was built, people at the time were young. Today, 30 years later, the second generation is moving out. It is natural cycle; no one gave it much thought but believed economic growth would continue forever. However, we know this is no longer the case. An aging society is one that consumes less: a family of 4 persons will eat 4 apples, but an elderly couple will eat only 1 apple. And reduced consumption means more vacant stores. With the work of refurbishing housing facilities, various projects are now going on to utilize the potential suburban housing complexes that are at a distance from the areas where the younger generation is living. Wakabadai is good example; "Community Office & Dining Haru" opened in 2014, catering to the residents' "job and food" interests. Another example is "Soramame," a place for children and their parents to freely get together. Both facilities are in the shopping street area. These are efforts to achieve "sustainable and cyclical community" by creating living spaces for the elderly as well as an environment that will attract members of the next generation to move into the housing complex. "Although it is still not easy at the moment, we would like to broaden our efforts to create a 'bottom-up development of sustainable town' in our housing estates."

図1 神奈川県の工業地帯（戦前を含む）
Fig.1 Industry Areas

図3 産業の空洞化
Fig.3 Hollowed-out of Industries (1991~)

図2 公社の団地
Fig.2 Housing estates of KPHS-Corp

図4 鉄道延伸と郊外団地開発
Fig.4 Housing development in the suburban areas

図5 神奈川の生活・産業・住環境の再構築にかかわる問題の整理
Fig.5 History of problems in the development of life, industry and housing

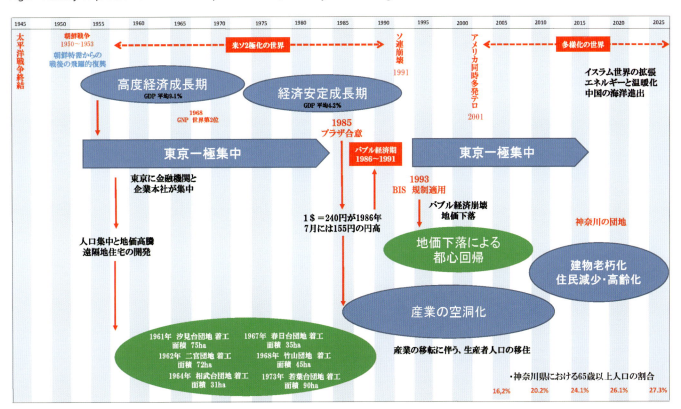

3-2 汐見台団地の計画と成長の経緯、これからの課題
Shiomidai Housing Estate Project – Overview of its History and the next Programs

節 健夫
SHITOMI, Takeo

■ 合わせ技で町名を統一

　神奈川県に会場を移した2日目。午前中は磯子区の汐見台団地でガイダンスとまち歩きを実施した。まずは神奈川県住宅供給公社の専務理事を務める節氏が、汐見台団地の開発の歴史を講演した。

　汐見台団地は磯子区の丘陵地帯（約75ha）にある大団地だ。計画概要は、団地面積74.7ha（22万6000坪）、3516戸、階数は3～10階でRC耐火構造。建蔽率は20％、容積率は50％。計画人口は約1万4000人だった。

　計画がスタートしたのは1959年春。当時建設が決まっていた根岸湾埋立臨海工業地帯の労働者向けの社宅として進出企業に街区単位で土地を譲渡する計画だった。今でいう「職住近接」である。（図1）

　土地買収は同年夏から交渉を開始した。このとき、「一団地の住宅経営」を選んだのは、土地収用の権限をもつことで公社主導の計画的なまちづくりを行ないたかったからだ。

　「地主が売ってくれないと民間の土地が残ってしまう。それは理想的なまちづくりを進めるためには厳しいと考えて、全面買収方式で始めた」と節氏は言う。

　また、当時は磯子区と南区にまたがっていたので、汐見台という町名に統一するために一団地の住宅経営と土地区画整理事業の合わせ技を使った。建設省（当時）からは「両方使うのは法の乱用だ」と批判を受けたそうだが、必要性を訴えて説得し、汐見台1～3丁目に統一した。

■ 時代を先取りした「電柱のないまちづくり」

　汐見台団地の団地計画は「近隣住区理論」をもとに、小学校を中心とした。またメインの通りに沿って、中央部に店舗、集会室、病院（当時は診療所）、広場を配置した。

　「公社は小さな団地設計はしていたけれど、インフラから提供する団地計画は初めてのこと。『理想的なまちづくりをやろう』とスタッフはかなり意気込んでいたそうだ」と振り返る。

　設計の基本指針は、（1）電柱のないまち（共同溝）、（2）道路は歩車道を分離、（3）住宅入口から直接車道に出ない、（4）パーキングエリアの充実、（5）樹木のある自然を残す、（6）土の移動は全て団地内処理──の6つ。歩車道を分離することは、当時はまだ少なかったし、本格的なモータリゼーションが起きていない時期にパーキングエリアの充実を図ったことも公社が未来を見据えた施策といえる。緑豊かな環境は、今では貴重な財産だ。

　指針のなかでもっとも難航したのは、電柱のないまちづくりである。「欧米では当たり前だけれど、当時はとにかく前例がなくて、電力会社との折衝がとても難しかった」と言う。

　電力会社からは「メンテナンス費用が予測不可能。戸数が確定しないうちに事前投資はできない」と言われた。そして「どうしてもやるなら、電線地中化工事費を全額公社で負担。さらに敷地内の引き込みケーブルはダブル配線にすること」という条件を提示され、公社はこれを受け入れた。約1億円の費用をかけて「電柱のないまちづくり」を実現した。

　共同溝式（下水道管を太くして配管スペースを確保）は事業スケジュールが一致しないうえ、法制上の制約も難しく断念。道路の両側に共同施設帯をつくり、歩道の下に配置した。（図2）

■ Back ground

Construction Started in Spring 1959
--Covering 75 ha. of Isogo hill area
--Built to provide housing for the workers of the Negishi industrial belt
--Primary research and design done by Kume Sekkei
--In July 1959, negotiation for land acquisition started
As a result of this negotiation, a city-planned road, which was supposed to traverse the property of Isogo Prince Hotel, was altered to go through our property.

■ Back ground

Construction Started in Spring 1959
--Covering 75 ha. of Isogo hill area
--Built to provide housing for the workers of the Negishi industrial belt
--Primary research and design done by Kume Sekkei
--In July 1959, negotiation for land acquisition started
As a result of this negotiation, a city-planned road, which was supposed to traverse the property of Isogo Prince Hotel, was altered to go through our property.

■ Issues faced in Housing Development and Management

Management of a housing complex/estate
  • Make regulations, e.g., number of rooms and floors, building-land ratio
  • What about facilities for the residents, e.g., supermarkets? Hospitals? Elementary schools?
Would any company buy a piece of land to build employee housing?
  • Property Area: 74.7 ha.
  • Number of rooms: 3,516
  • Number of floors: 3 to 10
  • Structure: RC (reinforced concrete)
  • Building coverage ratio: 20% (Gross)
  • Plot ratio: 50% (Gross)
  • Estimated population: 14,000 people
  • Population density: 187 per ha.
Developing Method
  • Urban planning: Management of a housing complex/estate
  • Decision of City Plans by Municipalities: March 11, 1960 /Project decided: August 25, 1960
  • Authorized: September 14, 1960
  • Construction Phases:
  • Foundation construction: December 1960–March 1962
  • Building construction: April 1962–March 1966
Basic Policy: Advanced Planning and Design (Fig.1)
  • Town without power/telephone poles
  • Separate car and pedestrian lanes
  • Separate building entrances for residents and driveways
  • Spacious parking
  • Landscape design
  • Use/recycle soil from the construction of the foundation

■住宅管理から「ニュータウンの経営」へ

　各企業がそれぞれに社宅を建てていくため、公社は「宅地法地面の肩に設けた斜面保護堤」「敷地の雨水排水の処理と芝張り」など設計協定をつくった。画期的だったのが「TV共同視聴アンテナの設置」だ。衛星テレビが普及しても景観が損なわれることはなかった。今も公社は原則的にこの手法を貫いている。

　公社はそれまで100～200戸程度の団地を扱っていたが、あくまでも「住宅管理」で道路や下水道などは既存インフラを利用していた。このような「ニュータウンの経営」は初めてだった。（図3）

　インフラ整備からまち全体を整備、経営するにあたり腐心したのは、継続的な費用をどう生み出すかだ。初期入居者に負担のないようにと考え、収益を期待できない公共施設と収益を上げる店舗を土地の原価に含ませて建設した。

　さらに、店舗の家賃収益で公共施設の運営を行ない、住民福祉のために使おうと「団地福祉協会」を設立した。

　また、外灯の維持や空き地の除草、中央下水の清掃など居住者も市も行なわない管理の維持が必要になるため、「団地共益費」を編み出す。土地所有権者は所有坪数に応じて負担し、住宅ができれば居住者の共益費とする。今も公社が管理している。

　こうして始まったものの、肝心の工場の埋立地進出が遅れ、企業の申込は予想以上に少なく、総戸数の2/3だった。公社としては大ピンチだ。そこで県と横浜市からの融資枠を拡大してもらい、余剰地を公社賃貸と分譲住宅に変更する。

　「結果的には社宅だけでなく、公社賃貸と分譲住宅があることでバラエティに富んだニュータウンになった」と語る。

　1962年に着工し、1963年に汐見台団地として初めての入居者を迎えた。（図2）

■若年層が多く、好ましい人口バランス

　時代とともに環境も経済状況も変わるので、1983年に「一団地の住宅経営」の計画を変更。（1）良好な居住環境を積極的に守る、（2）建物容積率等の規制に柔軟性をもたせる、（3）街区ごとの戸数を変えないとした。

　その後、生産拠点の海外移転などもあって、社宅から徐々に分譲マンションに切り替わっている（図4）。今は16区画ほどが分譲マンションだ。「一団地の住宅経営」の適用で街区ごとの戸数は決まっているため、一戸当たりの面積が大きい。面積が広すぎて売れない場合は共用スペースを広くとっている。「いずれにしても住宅街としてはよい方向に向かっている」と評価する。

　汐見台団地は、公社が管理する他の団地と違って、年少人口（0～14歳）の割合が高い。老年人口（65歳以上）の率も、多少上がってはいるもののかなり低い。横浜市や磯子区など周辺地域と比べても年少人口が多く、老年人口が少ない。これが汐見台団地の特長だ。（図5）「社宅が分譲マンションに切り替わっていったので現状はこうなっている。ただし、汐見台団地全体で見るとバランスがよいのだけれど、賃貸住宅は高齢者がかなり増えてきている」と分析する。

　「汐見台団地は、企業とともに職住近接のまちづくりを進めてきた。今となってみると不十分なところもあるが、ぜひ見ていただきたい」と締めくくった。

　講演後は汐見台団地のまち歩きを行ない、参加者は公社の職員の説明に耳を傾けた。（図6）

図1　立地企業と社宅（臨海部の企業の色は社宅の色と同じ）
Fig.1 Industry development in bay area and housing develop.

Design agreement of implementation planning
- Slope protection bank
- drainage system and planting lawn
- apply green fence for the site boundary
- One TV co-viewing antenna on each building (not each house) to maintain the scenery
- keep the same design for numbering each building
- safety measures during the construction

図2　電柱のない街路 / Fig.2 Street: power cable under ground

図3　「マンション管理」から「団地管理」に業務が拡大
Fig.3 Business developed from house to housing estate

Before the Shiomidai Housing Estate;
--Our main work was the maintenance of housing estates
--Design and construct housing complexes
--Using existing infrastructure
The Shiomidai project
--Managing the new towns
--Designing & building new infrastructure
--Operation and management of whole town
A system of managing a vast housing complexes
--Public facilities not expected to get revenue;
--Common room, Hospital, Kindergarten,

図4 社宅住宅用地を民間デベロッパーに
Fig.4 Large company employee housings estate go to private developers

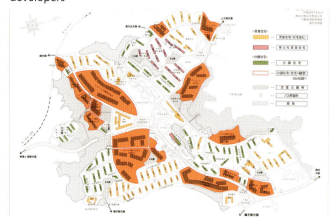

Children's library, Parks
Facilities that can be expected to produce revenue;
- Solution: include in the land purchase price
- Manage public facilities cost by Shops' revenue
- If any surplus, use this for residents' welfare.

Founded a "Welfare Association"

図5 汐見台団地と周辺自治体の年齢別人口割合
Fig.5 Age structure of populations Shiomidai and vicinities

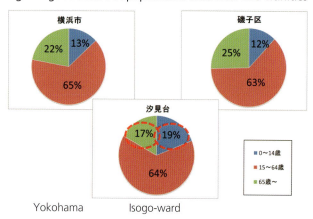

図6 汐見台団地「まち歩き」の様子
Fig.6 Shiomidai estate street scape

図7 汐見台団地の開発時の概観 / Fig.7 Over view in early development stage

3-3 神奈川における都市計画の発展と現状
Town Planning and its Issues of Kanagawa Area

鈴木 伸治

SUZUKI, Nobuharu

■産業構造の変化に応じた都市計画を

鈴木氏の専門は都市計画だ。特にアーバンデザインと呼ばれる地区計画を研究している。横浜市立大学に移ってからは対象分野を広げており、その知見のなかから横浜市および神奈川県のまちづくりの課題について言及した。

これまでの横浜の都市計画を振り返ると、明治末期以降は京浜臨海部の埋め立てが行なわれ、工場群の進出によって「工業都市化」が進む。戦後は、急速な東京のベッドタウン化によって郊外部で住宅地開発が行なわれた。

鈴木氏は「横浜の特徴として、人口の急増に対応した郊外開発が挙げられる。明治以降の人口と世帯数を見ると、横浜大空襲で多くの人が亡くなったが、1965年から1972年にかけては年平均9万6000人増加という事態だった」と語る。（図1）

当時の新聞を調べると、崖崩れが相次ぎ、都市型水害も起きていたという。「インフラ整備が十分ではないまま住宅地開発が進んだからだ。そこで横浜では宅地開発要綱をつくり、港北ニュータウンなどの開発に取り組んだ」と説明する。

現在、横浜が直面している課題として、（1）産業構造の変化と（2）人口減少・超高齢社会の到来がある。

まず産業だが、戦前からの素材型産業に加え、戦後は加工組立型産業の集積が進んだ。1990年代以降は、大規模工場の移転、閉鎖が相次ぐ。製造品出荷額は1990年のピーク時に比べ今は3割減。製造業事業所数、従業者数もほぼ半減している。（図2）

注目すべきは、従業者数のピークが1970年だったこと。人口が増えて住宅供給に追われているとき、すでに従業者は減っていたのだ。こうした背景から、多くの企業は工場や倉庫、社宅など資産を売り払い、それがマンションとして開発された。

現在、企業はかつての生産拠点から研究開発拠点へとシフトしているものの、その変化に都市計画が対応できていないことも大きな問題だ。

「工場のように見える建物でも、実は非常に高度な研究開発部門が入っていることが多い。しかし、工業専用地区なのでコンビニエンスストアやレストランが開けない。都市計画的な対応ができていないのはもったいない。用途変更を含めた規制緩和が必要だろう」と指摘する。

■目前に迫る「2025年問題」

一方の人口減少・超高齢社会の到来に対しては「横浜の都市計画は郊外部のビジョンが欠けている」と厳しく指摘。横浜市が発表した将来人口推計では、人口は緩やかに増えて2020年に約374万人でピークを迎えるが、2055年には約320万人まで減少する。たった35年間で54万人が減る。中核市あるいは政令指定都市が1つ丸々消えてしまう計算になる。

また、超高齢化という課題もある。団塊の世代が75歳以上の後期高齢者になる「2025年問題」は横浜も例外ではない。2025年には65歳以上の高齢者人口が約100万人に達し、75歳以上の後期高齢者人口が約61万人に達すると予測されている。また、独居高齢者も大幅な増加が予想される。

「日本は健康寿命と平均寿命に違いがある。日本人が寿命を迎える最後の10年ほどは健康上になんらかのトラブルを抱えながら生きていく

■ Recent history of Yokohama's town planning

Beginning in 1900, the "Keihin" (a part of Tokyo, Kawasaki, and Yokohama) coastal area had been expanded (reclaimed land) in order to fill the demand for factory space and to accommodate industrial urbanization. The neighboring towns were developed to house and entertain those employees who worked in the factories.

After WWII, Yokohama became a commuter town for people who worked in the Tokyo metropolitan area. Development of residential areas in the suburbs of Yokohama progressed rapidly. Although those residences also included company-owned houses/apartments for factory workers, they were mainly developed for "Yokohama-Tokyo" commuters. (Fig.1).

■ Two main issues facing Yokohama Town Planners

Yokohama faces many issues, and as regards town planning, the two serious concerns are a) a shifting industrial structure, and b) a declining population and an aging society. We need to consider countermeasures for these issues.

■ Shift of industrial structure (Fig.2)

The city of Yokohama's main industry before WWII was the handling of materials. But following WWII, it shifted to processing and assembling. After 1990, as large factories relocated or those businesses closed, the quantity of shipments declined by more than 30 percent, with the main purpose of business turning from manufacturing to R&D functions.

The number of offices/factories and employees is approximately 50 percent lower than in 1990. Therefore, those assets (housing complexes /dormitories /warehouses /factories) were sold by the ownership companies and then developed as apartment blocks.

In recent years, discussions have been held and studies made about how to redevelop those areas. However, many decisions, such as building a large logistics center, were made without incorporating the town planning ideas borne out of those studies. This result was that many decisions were unable to cope with the pace of change.

■ Population decline and ageing society (Fig.3)

In 2012, the city of Yokohama announced its population forecast for the coming years. It projects that the population will peak at approximately 374 mill. people in 2019 and decline to around 320 mill. in 2060. In 2025, the population of the "age 65 years old" category will reach almost 97 mill. and those turning 75 years or older will number 59 mill. At the same time, the number of people living alone will increase significantly. According to this estimate, in 2025 about 20 mill. elderly people will need assistance in their normal daily life, and around half of them will be suffering from some form of dementia. In Yokohama itself, the southern part of the city (the area of Yokohama farthest from Tokyo will experience a population decline and a progressively higher aging society. On the other hand, the northern part (area closest to Tokyo) will face the trend of an increasing population, thus creating an imbalance problem between North and South. Also, a number of residential developments in the suburb of Yokohama were built in communities

ケースが多い」と指摘。後期高齢者にかかる社会保障の経費は今後さらに増大していくと予測される。また、2025年には85歳以上の単身高齢者が4万人となるため、この対処も考えなければいけないし、2025年には要介護認定者数が20万人、認知症高齢者が10万人という予測もある。こうしたことをどう考え、いかに対応するのかは大きな問題だ。

■局所的に進行する「まだら高齢化」

1960年代から70年代にかけて、膨大な住宅が郊外につくられたため、入居時期によって特定の世代に偏る傾向がある。横浜市内では南部の区で人口減少と高齢化が進行しており、北部では人口増加の傾向が続く。鈴木氏が問題視するのは、こうしたことが「まだら」に進行することだ。

「横浜では郊外の住宅地開発によって、特定の世代が多いコミュニティが存在するので、人口減少および高齢化は局所的に進行する」と「まだら高齢化」と「まだら過疎」を懸念する。

「特に駅から遠く、丘陵地を開発した住宅地などでは、局所的な人口減少や高齢化が発生しやすい」と述べ、鉄道駅から1km圏内の住宅地を記した地図を投影した。(図3)

「オレンジ色の斜線がかかっているのは、鉄道駅から1kmの範囲。徒歩15分と見ていい。そのエリア外にも人口が多い。黄色い点は郊外の大規模団地・住宅地を指している」と説明。公営住宅や大規模な団地開発は駅から遠いところに立地していることが見て取れる。

別の地図では、交通の便のよくない郊外ほど高齢化が進んでいることがわかる。人口減少も同様だ。

■数十年後を見据えた長期的な視点

最後に都市経営上の課題とまちづくりの課題を整理した。

高齢化によって生ずる社会保障給付の増加は地方財政を確実に圧迫する。神奈川県下の自治体でもすでに財政的に厳しいところが出てきているし、今後はさらに負担が重くのしかかる。

「横浜市クラスの大都市ならまだしも、県西部や南部は人口減少・超高齢社会の到来は大きな課題となる」と語る。

カギとなるのは「健康寿命をどう延ばすか」だ。後期高齢者へのケアはセーフティネットとして必要なので、そういう状況に陥らないための取り組みを進めることで、多くの問題を回避できると鈴木氏は考えている。「要介護が1ランク下がると、数十億円というお金が生み出されるそうだ。日本の金融資産の50%以上は60代以上の人々が保有しているという数字もある。そういう人たちにどうやって楽しい老後を過ごしていただくか、郊外の生活の楽しさを満喫していただくかを真剣に考えなければいけない」と言う。

鈴木氏が勤める大学では、住宅地の空き店舗を活用して拠点運営をはじめたが、高齢者のもっぱらの関心は自身の健康のことだという。

「したがって、郊外型団地の再生を考えるには、いわゆる建築分野だけでなく、医療や福祉分野とも手を結ばなければ課題解決にはつながらないと思う」と述べた。

産業構造の変化への対応については、とにかく地域の雇用を生み出すことが必要だと指摘。「団地内で雇用を生み出せれば素晴らしい。さらに都心部や臨海部の産業用地をうまく活用する必要もある」と行政に迅速な対応を求めた。

郊外部の団地再生については、人口が減り続けることを忘れないでほしいと述べ、「整備などの公的支援に頼るだけでなく、数十年後にまた空き家ばかりにならないような、先を見据えた長期的な解決策が求められる」と締めくくった。

organized according to proper age groups. Therefore, the decline in population and a growing aged society will occur in those areas specifically, but this will not happen evenly throughout a larger area ("mottled" or unequally distributed aged society and depopulation map.) "Mottled" aged society and depopulation will occur especially in those areas located far from train stations or in hilly residential areas. Municipal's serviceability in this regard will not increase.

■ Economic issues of town planning (Fig.2)

Because there will be more senior citizens, local governments will face the economic burden of social security benefits (social welfare fees, senior citizens welfare fees, children's welfare fees, the general welfare to support daily living). The local governments will continue to bear this burden, which will inevitably lead to bankruptcy when the baby boomer generation turns 75 in 2025 if no effective changes are made.

The maintenance or rebuilding costs of the public facilities that were built during the period of rapid economic growth after WWII need to be considered, and this is also the reason why local government finances are in poor shape.

In order to stay on top of the shift in industrial structure, we need industries that create employment opportunities in those localities. We also need to seriously discuss replacing previous regulations that were based on supporting production. We need to correct infrastructure and environmental issues so that R&D-based organizations may thrive in these previous factory/industrial areas.

Urgent reorganization of multi-unit apartments in suburban areas is necessary; however, we must also consider not only the building (hardware) issues but also solutions for a society with a declining population.

図1 これまでの横浜の都市計画(人口急増と郊外開発)
Fig.1 Town planning of Yokohama - Population / Households
出典:「横浜経済の内発的発展」実態基礎調査報告書(横浜市経済局、2012)

図2　横浜市の製造業の時系列変化
Fig.2　Industry structure – change in Hokohama City
○ Business organization
▲ Total value of delivered products
● Total number of industry workers

図3　横浜市の郊外開発（オレンジ色は鉄道駅から 1km圏内。黄色い点は郊外の大規模団地・住宅地）
Fig. 3 View on the housing estate development in Yokohama
■ Housing estates　— Railway lines

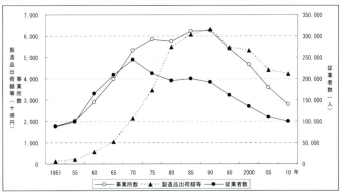

図表1－10　横浜市の製造業事業所数・従業者数・製造品出荷額等の推移

出所：横浜市「横浜市の工業」（4人以上）

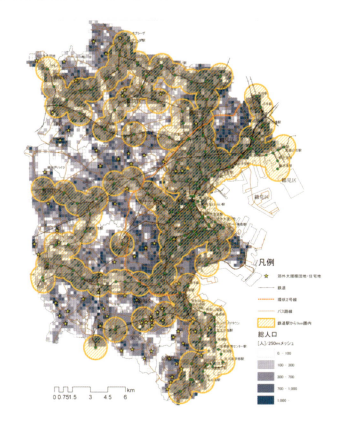

図4　横浜港　1930年ころ
Fig.4　The port of Yokohama, 1930

077

3-4 討議とまとめ -
現場を見てポジティブに考える、コンパクトシティの意味、国際交流のあり方
Discussion and Conclusion – Positive thinking on site, Compact City prospect, International exchange

原 /HARA、猪股 /INOMATA、シュトレーブ /STRAEB、大村 /OMURA、鈴木 /SUZUKI、永松 /NAGAMATSU、澤田 /SAWADA

■「現場」を見てポジティブに考える

原 汐見台団地の視察と公社のサステナブル社会のまちづくりの取組み紹介から日独間の意見交換の可能性を自分は感じた。この点どうか？

大村 日独は、これから脱産業化社会を形成して行かなければならないという点で共通な立場にある。工場跡地を収益性の高い土地利用に転じるのは容易でなく、コストをかけずに"人を惹きつけられる"方策は何かを考えている状況だ。
　土地利用では、ドイツは郊外地域の開発規制の厳しさが違う。
　ゲーラ市の場合、40haの自然の土地を開発するには20haの代償を用意しなければならない。その土地が農地などすでに開発された土地であってはいけないと定めている。両国を比べると"サステナブル社会づくり"の面で進捗度の違いが大きい。
　一方、日本における工場跡地利用のアウトレットモールや大型物流センター開発では、通常3年から5年で開発投資の回収を考えたプランニングになっている。それは"次世代の産業廃棄物的土地"を計画するのと似ていて、サステナブルにならない。

猪俣 ドイツの旧東独エリアでは25年前のベルリンの壁崩壊後に、3年から5年でコミュニティが急に"歯抜け"になった。ここ神奈川ではゆっくりと"歯抜け"が進行している。そう考えて旧東独のサステナブルなまちづくりの実状を学ぼうと考えた。東京会議では富山市も紹介されたが、神奈川とは距離があり過ぎるし自然環境も違う。なお、今日の横浜会議は、ここで"結論"の出るのを予定していない。むしろ交流の始まりと考えている。
　ちなみに、公社住宅の一つに戸建住宅800戸という事例・二宮団地がある。現在300戸が空き家で、人を呼ぶ手だてを検討中だ。二宮は軍需産業の盛んだった時代に社宅として開発され、声掛けする人たちは周辺に見当たらない。二宮について、300戸をつぶして500戸を高品質住宅化することを検討したが、今の公社の経営としては、800戸の住宅資産保有し続ける方針だ。

大村 ミラーさんから聞いたゲーラ市のこと。市の南に4万人の団地、北部に2万人の団地があったが、壁が崩壊して南の団地の住民が2.2万人に減った。減築して戸数を減らし、リノベーションや外部環境を整備したら住民の満足度が上がったという。もっとも旧東独と神奈川では、住民の年齢構成や人種構成が異なり、失業率はゲーラが高いし、麻薬の問題もあって周辺住民に不安感を持つ人も多く、日本とは異なる面もある。

猪俣 二宮団地の近所に別の公社住宅がある。自分の理事長就任時には、景色は良い団地だが遠いので視察に行かれなかった。ところが、その団地の5棟のそれぞれの空室は30戸の内2、3戸に過ぎない。東京から来た65歳以上の一人暮らしのようだ。家賃が4万円だから週末は東京の自宅に帰っているかと思う。二宮住宅には、団塊の世代をまとめておよびする計画も検討している。

原 公社の相武台団地には"二戸持ちさん"が100名もいるそうだ。家が手狭になったとか、子世代が遊びに来た時使いたいニーズがあるよう

■ Think positively of the "work place"

HARA I was hoping for the possibility to exchange opinions regarding the Shiomidai residential complex and the session on introduction to sustainable Machidukuri by Mr. Straeb.

OMURA Japan and Germany have in common that they both must enter a deindustrialization phase. It is not easy to turn industrial land into a high- profit land. Both countries are currently researching ways to attract people to these types of sites at a low-cost investment.
　In Gera 20ha of non-artificial terrain must be left unbuilt when planning for a construction project of 40ha. However in Japan large distribution complexes which are built in brownfield, have planning schemes that normally look for a return on the initial investment during a 3-5 year period.
　This is a planning concept that will create "industrial wastelands" for future generations, and it is far from being a sustainable one. The difference in progression between both counties is apparent once progress in "sustainable society development" is compared.

INOMATA After the fall of the Berlin Wall, the former East Germany recovered from a sudden community collapse in a 3-5 year period; in Kanagawa, however, the collapse was more gradual. I wanted to use this opportunity to learn about the present condition of the former East German sustainable Machidukuri project. It should be noted that I do not predict a "conclusion" will be reached from this Yokohama comference, though I do believe that it will serve as the beginning of a relationship for a more sustainable future.
　I would also like to talk about the Ninomiya housing estate case. This estate is a corporate housing project with a total of 800 detached houses that were developed during the military-industrial era. This would be one of good example for our realistic discussion.

OMURA As Mr Miller mentioned previously, the city of Gera had large housing estates, one in the north and another in the south. When the Berlin wall fell, 40000 inhabitants in the south decreased suddenly to 18000 due the migration to the west. As a result of this radical change in population Gera decided to demolish a number of its housings and renovate on the remaining residential buildings, this translated to the rise in satisfaction among its citizens. But it should also be noted that the unemployment rate Gera is far higher in comparison to Kanagawa and this has translated into the high crime rate and discomfort of its citizens.

INOMATA There is other housing corporation in the same district as the Ninomiya housing estate. Currently, there are only 2 or 3 vacant units out of the 30-unit dwellings. Most of the people who have moved in seem to be singles over 65 and to be living in Tokyo on weekends. After observing this phenomenon, the Ninomiya housing estate started to come up with a plan to attract people from the baby boom generation (people over 65).

HARA There have been reports from the Sobudai-Danchi residential estate indicating that there are more than 100 residents who have a "second home." Upon further investigation, the residents have revealed that many people required more space to live or that they wanted a house for when younger generations came to visit. The residential estate has strong community ties and frequently holds activities such as fireworks displays or gardening.

だ。あそこの団地コミュニティは意外とまとまりがあり、農園をやったり花火大会をやっている。東京の事例ではメゾン・アオキ・マンションが共用部分を充実させて若手住民を上手く獲得している。

東京会議の"まちづくり建設産業の変革ー「場」の産業へ"では、松村先生が「あり余る空間資源をポジティブにとらえて、再生・活用しよう」と呼びかけていた。神奈川は東京から遠いが中心市街地ならこうしたプロジェクトの可能性がありそうだが？

鈴木 横浜の場合、横浜駅周りとみなとみらい地区の再開発しか動いていない。他の場所の可能性は見当たらないと思う。

空き家住宅など遊休環境資産をポジティブに捉えようというメッセージはシュトレーブさん、ミラーさんからも度々あった。自分も松村さんのリノベーションスクールに参加してポジティブな熱気に触れてみた。その議論では、建築基準法という壁を突破できないか？という意見も聞いたが、神奈川ではどうか？

原 グレーゾーンの多い基準法については、神奈川でも緩和の環境が整備されて、これからは自治体が動かして行く段階だ。

■コンパクトシティ再考

シュトレープ この10年のドイツでのプロジェクトを概観すると、多くが都市中心部を対象にしている。そうした"まち"で伝統的な空間構造を継承しようとすると、簡単にはコンパクト化できないケースもある。

大村 東京会議では、コンパクトは必ずしも物理的なコンパクトを差さず、むしろコミュニティのあり方を重視するということだった。今日は、福祉や高齢者課題の対応について現場を見た。これにドイツ事例でのエネルギー、モビリティを加えてコンパクト化することになるだろう。

澤田 コンパクトシティとは、そのシティの人々の生活が"地産地消"の意味を含めて循環型、自律的に運営管理されている空間単位だと思う。

■日独交流研究の今後

永松 IBA方式は、時代の発展につれてテーマも変化している。しかしイノベーション支援策として、良いアイデアを提出し、採択されれば実際に建設して、その良さを検証する方式であり、"開いていく意識"が背景にある。これは、閉そく感や「壁」の存在する日本にとって参考になると思った。

原 自分の住む大磯町を含めて、グローバルな観点での位置づけが把握できた。大磯は開発投資のネタに乏しいが楽しさがある。大磯のまちづくりの進め方についても、このシンポジウムからそれで良いのだという、確認ができたと思う。

all though there is little development in the town where I live, I feel more positive about the Machidukuri possibilities available.

In the Tokyo conference speech, Professor Matsumura stated that "we should positively embrace the spatial resources and use them for the renovation process." Kanagawa is too far from Tokyo but in the central districts it may be possible to accomplish a project of this sort.

SUZUKI In the case of Yokohama, the redevelopment process has been limited to the Minato Mirai district and Yokohama station, and I do not see any other regions that can be redeveloped.

HARA After attending the symposium, I feel more positive about the matter of renovation and have come to the conclusion that in order to break the wall created by the current building regulations, a new form of building regulations might be required.

The current standards have many grey areas, and the local government of Kanagawa has started planning ways to relax many of its standards

■ The regeneration of compact cities

STRAEB When looking back at the German projects of the past 10 years, we see that the majority of them have been carried out in the city centre. The problem lies in preserving these traditional "city centers" during the compact city process.

OMURA In the Tokyo symposium, it was determined that the term "compact" described the social compactness of the community rather than the physical compactness. This symposium has introduced measures to cope with welfare and aging populations; the next step would be to add the energy and mobility measures like those introduced by German people.

SAWADA compact city is characterized by the lifestyles of its citizens, and I think that in order to be a compact city, it must allow the residents to have a circulation economy that allows for "localized production and consumption."

This circulation economy must also be supported by a "areal unit" that allows for an autonomously operated management system.

■ Future Japanese and German research exchange

NAGAMATSU The IBA scheme has been changing to adapt to the passing of time. The scheme has an innovation support system that allows ideas to be tested and evaluated, but most importantly it has an "open perspective." I have felt that this will serve as a good reference for country with many "social barriers" such as Japan.

HARA After attending this symposium, I feel that I have managed to gain a more global perspective of the Machidukuri process and all though there is little development in the town where I live, I feel more positive about the Machidukuri possibilities available.

団地再生プロジェクトの新ツール 住民参加マスタープランづくりの地形模型 FaBG ISシステム
Shiomidai Housing Estate Project – Overview of its History and the next Programs

斜面の男山団地（京都）／Typical hillside housing estate, Otokoyama, Kyoto

横浜市旭区は、帷子川源流域のアップダウンの激しい地形に位置し、1960年代から団地や宅地開発が広範囲に行われてきた。ところが2003年には人口25.4万人でピークを迎え、その後は人口減少と急激な高齢化が進み、2035年には人口が2割減少し、高齢化率は37.8%となることが想定され、都市計画マスタープランを改定することになった。この改訂に当たって、市民が身近な生活圏について考えるためのツールとしてFabGISが開発された。

FaBGISは立体的な地形模型上に、各種施設の立地や将来の居住者数などの統計データ、地理データを投影するシステムである。今までは、地図やモニターで2次元的に表現されたデータが、住民生活の変化と合わせて立体的に表現できるシステムであり、ワークショップ・ツールとしては、前例のないものである。

日本には複雑な地形の場所に開発された住宅団地が少なくない。FaBGISはそうした団地の再生でも役立つコミュニケーション・ツールである。

Asahi Ward of Yokohama is located along with Katabira–river in a complicated landform place. Large housing developments started there in 1960s. Population of Asahi Ward peaked in 2003 at 254,000. The decline thereafter has been rapidly, and it's aging is expected 37.8%.

"FabGIS" is a communication tool for the planning re-development of the existing housing estates that enables citizen's discussion about "before-situation and after-situation" of their living built-environment.

FabGIS a landform modeling system combined with projection of GIS-data on it. Future population data and community facility condition can be projected on the three-dimensional model. FabGIS makes "before–after" situation visualize and supports the discussion in a more comprehensive manner. FabGIS system is a first communication tool for citizen–participation to workshop and expected to be applied in Machidukuri process further in other housing redevelopment projects.

整備されたセンター／Center area after redevelopment

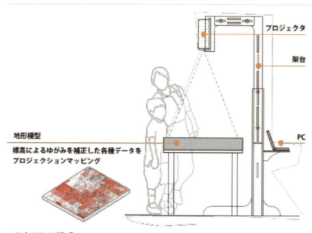

FabGISの構成

FabGISを活用したワークショップの様子

FaBGIS: Fab+Geographic Information System
開発者： 庵原悠（(株)岡村製作所）、秋吉浩気、深井千尋（慶應義塾大学）
片岡公一（(株)山手総合計画研究所）
http://www.y-p-c.co.jp/ypc/

FabGIS: Fabrication + Geographic Information System
開発：ファブ・シティー・コンソーシアム
Development: Fab City Consortium、http://fabcity.sfc.keio.ac.jp/
庵原悠（岡村製作所）／Yuu Ihara (Okamura Corporation)
秋吉浩気（慶應義塾大学）／Koki Akiyoshi (Keio University)
深井千尋（慶應義塾大学）／Chihiro Fukai (Keio University)
片岡公一（山手総合計画研究所）／Kimikazu Kataoka (Yamate Planning Cabin)

第4章　関西・滋賀県で考えるサステナブルなまちづくり
Chapter4　Sustainable Development of Shiga – Community

琵琶湖／Biwa Lake (the largest lake in Japan)

滋賀会議 Shiga Conference

仁連孝昭（文責：仁連）　滋賀のサステナブル社会のまちづくりのあり方
Prof. NIREN　Community Development for Sustainable Shiga

松岡拓公雄（文責：松岡）　ドイツのコンバージョン、リノベーションの建築デザイン
Mr. MATSUOKA　Learning from Conversion and Renovation Design in Germany

討議とまとめ—まちづくりプロジェクトの始め方、地域コミュニティに特有の課題
Discussion and Conclusion

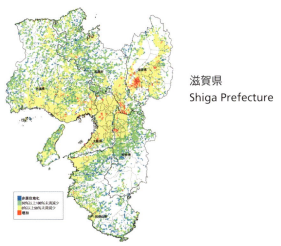

滋賀県
Shiga Prefecture

関西圏と滋賀県／Kansai Metropolis and Shiga Prefecture

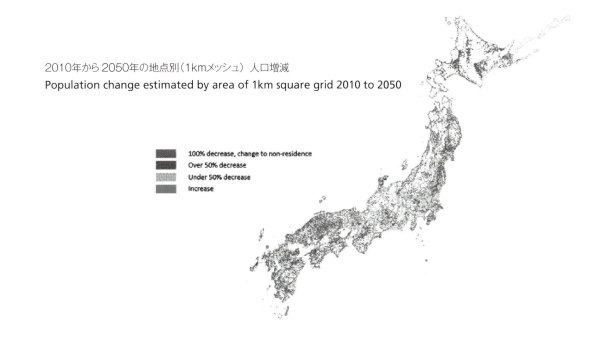

2010年から2050年の地点別（1kmメッシュ）人口増減
Population change estimated by area of 1km square grid 2010 to 2050

100% decrease, change to non-residence
Over 50% decrease
Under 50% decrease
Increase

4-1 滋賀のサステナブル社会のまちづくりのあり方
Community Development for Sustainable Shiga

仁連 孝明
NIREN, Takaaki

■ 人口減少を考える

　滋賀県を閉鎖地域と見て、人口減少の動きを人口動態、出生率、死亡率からその傾向を測ると図のようになる(1981-2011年)。これに年齢別生存率・死亡率のデータを加えさらに「日本の人口推移」予測に「滋賀県のデータ」を重ねて見る必要がある。日本では2008年人口増加のピーク(1億2800万人)となったが、2050年には9700万人まで減少している。2010年時から見て24%もの減少でそのスピードが速い。また年齢別人口構成を見ると、老年人口(65歳以上)、生産年齢人口(15-64歳)、年少人口(15歳以下)に区分し、高齢者割合の急速な増加と少子化の動向の顕著なことがわかる。これが人口動態の現状である。(表1、図1)

■ 滋賀県の人口動態

　ところが滋賀は開かれた地域であり、この人口増減の背景には、地方から大都市への人口移動、大都市から大都市周辺への人口移動、東京への一極集中という人口流動があることを含めなければならない。
　図は近畿各県および東京都の社会増減人口率を示す。5年後の人口を母数として5年間の流入人口から流出人口を控除した社会増減人口の割合を%表示した。
　Aの矢印で示した期間にわたって大阪、東京などの大都市に人口が移動している事を示し、Bは同じように大都市の周辺、すなわち近畿では滋賀、奈良へ人口が移動している期間を示している。(表2)
　ちょうどこの期間には大都市である大阪、東京では人口流出が起きている。Cは大都市周辺への人口流入が止まり、東京への人口の一極流入が始まった事を示している。
　さらに2050年の人口増減状況が国土交通省「国土のグランドデザイン－1kmメッシュデータ」から読み取れた。
　2001年から2050年の人口の増減について示され図などがあり、50%以上減少すると予想される地点が図2に表示されている。また、2050年までに無居住化する地点、10%の減少・増加の予測される地点を示す図もあるので、県下の各市の将来状況がより具体的にわかる。この人口動態と人口ピラミッド(1970年、2010年、2030年、2050年の推移)とから滋賀県という"広域コミュニティ"がどのような社会に変化するかが具体的にわかる。(図4を参照)

■ 人口成長社会から人口減少社会への転換とは何か？

人口の成長期と減少期とでは社会計画の課題が異なる
○人口成長期の問題とは；
・増加する人口による追加的ニーズに対する供給不足という問題
・すなわち住宅の供給、インフラストラクチャの追加的整備(道路、上水・下水道、学校)である。
○人口減少期の問題とは；
・遊休資産(未利用土地・建物)の増加によるコミュニティ、都市、地域全体の質の劣化であり、
・人口成長期に構築した環境資産である土地・施設、インフラの維持サービス供給システムであり、

■ Present Situation of Shiga

　Shiga lies at the center of Japan, linking the great cities of Tokyo, Nagoya, and Osaka. Shiga is also the nodal point which connects the land on the Pacific side of Japan's archipelago to the Sea of Japan side. With this, Shiga has become the location for the concentration processing and assembly factories which in turn distribute materials and products both to and from other regions, which promoting housing development in the southern area of Shiga for the neighboring Osaka metropolitan regions.
　As a result, with this increase in manufacturing, Shiga has also experienced a rapid increase in population. The population has grown from 842,000 in 1960 to 1,416,000 in 2013, nearly doubling in just over 50 years. However this trend has already begun to normalize, and even begun to decline. According to the estimate of the Ministry of Land, Infrastructure, Transport, and Tourism, the population in Shiga will have decreased by 13% over the 40 years from 2010 to 2050 and will be approximately 1,227,000 by 2050. In comparison to the total population of Japan, which is estimated to decline 24% over 40 years, this reduction rate seems somewhat less severe. But this 13% population decline appears quite differently in terms of space. In an estimated square kilometer grid of a residential area, it is calculated that 28% of it will see a reduction of more than half of the population; 65% will see less than half reduced, and only 7% will see an increase. (Tab 1, Fig.1)

■ Problems of regional planning

　The population decline will likely bring about the following problems;

1) A withdrawal of basic and necessary services such as commerce, social welfare, medical care, sanitation, and security.

2) An increase of vacant homes, non-operating facilities, deserted cultivated land, abandoned forest land.

3) Inefficient operation and maintenance of infrastructure such as roads and water supply and sewage systems.

4) Loss of regional employment opportunities and a decrease in the working-age population ratio.

　There is deep concern that these conditions will cause a diminishment of the local community. This situation can nevertheless be seen as an opportunity to develop sustainable community and to solve these impending local problems. Shiga is the nodal point of the Japanese islands and the watershed for Lake Biwa having 670 square kilometers with a surface area of 63 kilometers extending from North to South.
　The population growth and the changes within the watershed have made a huge impact on Lake Biwa and the surrounding ecosystems. As it were, this unsustainable lifestyle is reflects directly on the lake. Forestlands have been abandoned and therefore neglected in the change from forestry products to industrial goods; wetlands have been altered by agricultural land reclamation and improvement; sprawling development on the one side and hollowing out of urban

それらが個別的に構築・運営管理されて、経年劣化しているので、その包括的な扱いのための「全体像」が現在求められている。それは単なる「空き家」対策では対応できない。全体的なビジョンが必要である。そうしたことから、人口減少社会でのプランニング・コンストラクション・マネジメントのあり方の基本として、開発・用途転換（コンバージョン）、そして保存をバランスさせたコミュニティ、都市、地域の全体ビジョンをつくる必要があり、それを早急に「具体的かたち」にすることが求められている。

■人口成長社会から人口減少社会への転換の意味するもの

我が国の人口成長期は戦後に始まり1970年代に「豊かさの時代」を迎えた。そのことを踏まえて「サステナブル社会のまちづくり」を計画するに当たっては、「メガトレンド」を考慮しなければならない。
メガトレンドとは、「気候変動」と「政府の公共支出能力の弱化」であり、地球規模の自然界の大変化であり、後者はバブル崩壊後現在まで続く国家運営の低迷の事態である。
（注：この事態はチューリンゲンにおいても同様である。）

■サステナブルな地域づくりのビジョン

表2-aの上欄に人口成長期の問題と人口減少期の問題を示した。表の中央に人口減少社会への転換の「基本原理」を次のようにまとめた。
○人口変化（とりわけ減少）に対応できる"柔軟な地域づくり"
○気候変動に対応し、自然への負荷を抑える"省エネルギー・省資源"の循環型社会づくり
○公共負担を抑制する事業の仕組み
（解説）
＊「空き家・空き地の活用」とは、それら遊休化したものの、市街地内農地・緑地、保育・ケア施設、サービス施設への転用を図ること。
＊ネットワーク型インフラの分散型インフラへの転換とは、道路、上下水道、エネルギー、通信などネットワーク型の中でも「分散型下水道処理の導入」が重要であること。
＊地方の資源利用と地方の雇用機会の創出とは、食糧に地産地消（生産者がその場所で自ら消費する）、および、地域資源活用型観光・ものづくりを最も重要と考えること。
＊分散型再生可能エネルギーの利用とは、滋賀地域の全体像を対象とし、森林の活用（コンポストなど）をはかること。（表2、表3）

■遊休資産の管理

このビジョンは、滋賀地域のまちづくりのプロセスの初期に提示し、それに基づく関係者の議論を経て実施計画に進むためのものである。そのためこのビジョンの内容は「社会システム」の提案であり、今後の議論では、「遊休資産の管理」に関する一定の展望を持つことになる。
ここで言う「遊休資産」とは、森林・河川・湖沼などの自然環境資産に始まり、地域全体に広がる様々な用途の土地、インフラストラクチャー、都市・コミュニティ内の住宅、諸施設などの内の現在は環境資産として使用されていない遊休状態のものを差す。
遊休資産の管理については、その資産へのニーズがあれば、資産の持ち手と利用希望者の間で取引が成立するので、余計な規制さえなければその資産はうまく利用できる（表3-a左上）。例えば町家をそのまま修復・改修しレストランに利用すること。しかし持ち手が明確でなければ、持ち手が資産を利用することはなく、また利用者を探すことなく放置するならば、取引の当事者の一方がいないので取引は成立しない（表3-a右上）。利用者のニーズがあるにもかかわらず、都市に放置された空き家、放置された農地などがそれに当たる。そのままでは利用

area on the other has caused a disruption in the ecological balance, accelerated by the societal mass-consumption. (Tab.2 , Tab.3)

Although we are at the stage of a population peak, there are still unused agricultural lands, farmlands, and urban area. In spite of this, these lands are being left idle while there continues to be development of new properties. Thus this development pattern brings a high environmental impact and is also inefficient. The reasons behind this are;

1) The institution of land property and use inhibits changeover to users and for converting land use, and

2) The zoning system, which lays down negative land use and no active land-use plans, bringing about the objection of the community.

In order to change this situation, we need to draw up on a community land-use vision which will give the perspective of the sustainable community and then have to the power to implement it.

表1　人口減少ー出生率の変化
Tab.1 About population decline – birth rate change

	1981	1986	1991	1996	2001	2006	2011
母の年齢コーホート	1.74	1.72	1.53	1.43	1.33	1.32	1.39
15-19	0.0196	0.0196	0.0188	0.0188	0.0289	0.0250	0.0227
20-24	0.3697	0.3016	0.2244	0.1988	0.1980	0.1871	0.1710
25-29	0.9074	0.8557	0.6953	0.5631	0.4782	0.4353	0.4349
30-34	0.3669	0.4473	0.4722	0.4895	0.4425	0.4516	0.4836
35-39	0.0693	0.0891	0.1115	0.1395	0.1659	0.1886	0.2390
40-44	0.0082	0.0094	0.0118	0.0155	0.0199	0.0286	0.0408
45-49	0.0003	0.0003	0.0003	0.0004	0.0005	0.0007	0.0011

図1　人口減少ー男性と女性
Fig.1 About population decline – male & female

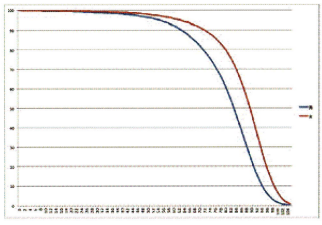

平成25年(2013)簡易生命表(厚生労働省)より作成

者のニーズが存在しない(表3-a左下と右下)。これら資産が利用されないまま放置されることは、遊休している資産そのものだけでなく、それが位置する都市、農地、森林全体の資本としての価値を損なうことになる。したがって、これらの遊休資産については用途転換と持ち手、利用者転換を含む総合的な対応が必要となる。

図2　2010年比 50%以上の人口減少の見込まれる地域
Fig.2　Area of population decline by 50%: 2010-2050

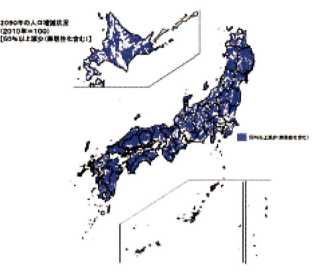

表2　人口減少―生産年齢猪人口の減少
Tab.2　About population decline – "productive population"

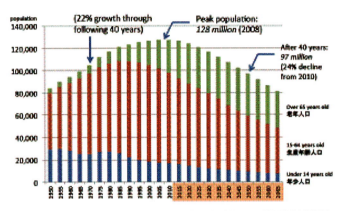

表2-a　サステナブル地域づくりのビジョン

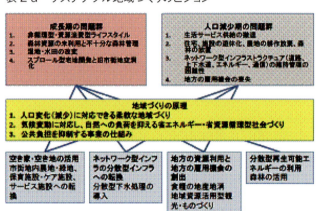

Tab.2-b　Vision for the sustainable community

表3　滋賀の人口動態
Tab.3　Population Change in Shiga
　A: 大阪、東京圏への人口移動 / Flow to Osaka & Tokyo
　B: 大都市周辺での移動 / Move in suburbs of big cities
　C: 東京圏内での移動(郊外へ) / Move to suburb, Tokyo

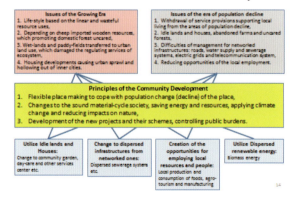

表3-a　遊休人工環境資産の管理

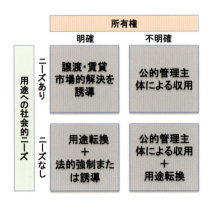

Tab.3-b　Management of the idle capital

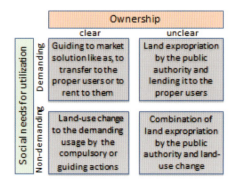

4-2　ドイツのコンバージョン、リノベーションの建築デザイン
Learning from Conversion and Renovation Design in Germany

松岡拓公雄

MATSUOKA, Takeo

　環境先進国と言われる日本は何故、未だにドイツを中心としたEUに学ばなければいけないのか。日本の都市計画や都市政策は、住民の安全、安心、快適に住む環境を保全するという名目はあるが、背景には経済優先があり、どうしても企業営利を優先した建設促進が推進されたしくみになっている。一方、ドイツはドイツ憲法で保障されている「何人も人間の尊厳に値する生活をする権利があり、それを担保する義務が行政にある」という市民の住む権利を守ることが優先される。

　日本の事情は全く逆である。都市計画の最も基本的な部分、人が中心であることがおざなりにされ、計画のための計画が進められてきた。戦後の「大量生産・大量消費・大量廃棄型社会」から「循環型社会」への転換が先進的に進んでいるドイツだが、同じ路線を目指す日本との基本的な違いは人間の尊厳に関わる部分である。環境先進国と言われるドイツからみると同じ目標を目指しているようでありながら、日本の「循環型社会」への進め方は違う方向にある。ここに学ぶべきことが見えてくる。

ドイツの「循環型社会」とは何か。建築で言えば、
○必要最小限のものを造る、
○資材選定や断熱など省エネ基準を厳格に守らせ、
○完成したものは長期間使用する、
○最終的に解体した後も、その部材を回収し再生や自然に還していく、
というルールになっている。

　国民の手によって、また国民の代表である政府によって、きめ細かいルールの下にシステムが構築され、しっかりと管理されている。それが建築における「循環型社会」である。

　日本でも「環境建築憲章」が2000年に日本建築家協会ほか各団体から掲げられ、長寿命・自然共生・省エネルギー・省資源／循環・継承の5項目が謳われているが、実践における制度が未だに確立していない。

　建築の場合、建築がより長く使われることが「循環型社会」に大きく寄与する。その手法としてコンバージョンやリノベーションがある。その中で最も比重を占めるのが戸建て住宅や集合住宅、我々の住まいであるが、その先行事例が特に進んでいるのがドイツなど EUである。その手法のひとつで様々な「減築」の考え方も、単なる空き屋対策という一面もあるにしろ、管理の面から求められていることの意味合いは大きい。

　我々は都市計画で高く評価されたラインハルト市長率いるドイツのライネフェルデ市の集合住宅の再構築の実践や、ルール地方の広大な範囲に広がる工業地帯産業施設の負の遺産を変身させ経済の活性化、地域の再生に成功したIBAエムシャーパークプロジェクトなどにその先進的事例を見ることができる。

　Why should a country as environmentally advanced as Japan need to learn from Germany? There is a pretense here of urban planning and policy to preserve a living environment where residents can live comfortably in safety and with peace of mind. Nevertheless, in the background there is a determination to give priority to corporate profit when construction activities are being promoted.

　On the one hand, Germany's constitution guarantees "…a right for all people to live in dignity, and it is the obligation of the state to secure this right," a guarantee that makes a citizen's right to life a priority. The situation in Japan is just the opposite. There is apathy toward the most basic element of urban planning, which is to focus on people, and planning seems to be done for the sake of planning. In contrast, Germany's postwar mass producing, mass consuming, and wasteful society has transformed into an advanced "recycling-based society" of the future. The basic difference is that Japan, which is striving for the same goal, is lacking the element of human dignity. From the perspective of environmentally developed Germany, Japan does appear to be seeking the same objective, yet its method of advancing toward a "recycling-based society" is heading in a different direction. It is here that we can see what we can learn.

　So what characterizes Germany's "recycling-based society"?

　In terms of construction: 1. Build the bare minimum. 2. Select materials and insulation that meet strict energy-saving standards. 3. Completed projects should have an extended period of use. 4. Following demolition, debris should be collected and recycled, or returned to nature. 5. The people or the government, which represents the people, should carefully create a system that is managed according to a detailed set of rules.

　This is a "recycling-based society" as it pertains to architecture. In 2000, the "Charter for Environmental Architecture" was published by the Japan Institute of Architects and other groups. Long service life, natural symbiosis, energy conservation, and resource conservation / circulation—these 4 items in succession are declared in the charter, but a system for implementing them has yet to be established.

　Building structures that have longer use periods will contribute significantly to a "recycling-based society." Following this line of thinking, we can utilize conversion and renovation.

　Also along this line, we can see the development of single-family housing into cluster housing. These are examples of the kind of housing the Germans and also people in the EU generally are living in. Through this method, there are various ways of thinking about reducing new construction. It is a simple way to decrease the number of vacant homes. Even this one aspect alone holds important implications from a management perspective. We can see examples of these types of advancements in the highly acclaimed cluster housing project in the city of Leinefelde, which was spearheaded by Mayer Reinhart; or in the success of the IBA's Emscher Park project, which reclaimed the region from the negative impact of the industrial facilities / brownfield on the natural environment in the Ruhr district.

　Moreover, with Japan's declining population, the social issues surrounding the dwindling birthrate and aging population have come into view. This is not a new issue; it's a phenomenon that was predicted long ago. There are vacant houses, vacant shops. And another problem seen particularly in Shiga Prefecture, which has more temples than anywhere else in Japan, temples are being abandoned as the number of supporting families declines. The alarm is sounding for the preservation of our traditional cityscapes and culture that are part of our history. While cities and communities are starting to become more compact, preservation is a huge issue that the central and local governments must seriously tackle together. While the policy of cities and communities are to become more corporative and also integrative, and they have to carry out their tasks more efficiently.

手前はバルコニーの増設、
突当りは壁面の多様化

様々なバルコニーを付けて中庭との対話を演出

住棟のすき間を屋根でつなぎ"ゲート"空間を作る

アーバン・ビラ（長大住棟の減築）

ワイマールのバウハウス大学の図書館と講義室（間を埋める）

ポツダムの住宅＋オフィス（ファサードを保存）

出典
Neubauland Exhibition, Deutsches Architekturmuseum, 2007

086

4-3 討議とまとめ—
まちづくりプロジェクトの始め方、地域コミュニティに特有の課題
Discussion and Conclusion

■討議

仁連 まず、多様な主体の参画を得て、計画をつくり審査をし、実行する一連のIBAという都市計画の手法について、成功の秘訣について補足いただけないか?

シュトレーブ 人口縮小社会という課題を共有することからスタートし、国・地域レベルの政治を含めて利害関係者全員で話し合うことが重要。そうすれば、関係者間の協力によって様々な利益が得られることも伝えられる。

仁連 まず課題を関係者が共有するステップがある。このステップが日本の場合はプランニングの中に含まれていないことが多い。政府や自治体による課題の提示から始まることが多い。これがドイツとの違いではないか?

参加者 ドイツでは脱原発が進んでいるが、日本では進まない。日本の裁判制度と世論形成が原因ではないか?日本では廃炉技術が進んでいないことも原因だ。
　多様な主体がそれぞれの役割を認識して協働して行く必要があることが解った。それが日本の地域での会合ではあまり進んでいない。ドイツとの違いは何か?

シュトレーブ 社会を構成するそれぞれの主体には役割がある。まちづくり活動は政治的なものに留まるものではない。会合に参加する各自は役割に応じて、できることとできないことの区別を明確にし、自身の役割を果たすことに努力する。そうした会合で重要なのは全員が学び合える状態をつくることだ。

松岡 日本の自治体に設置する委員会の場で問題になるのは、住民代表が地域コミュニティの人間関係や力関係に縛られる点だ。そのために公正な意見が出せないことがある。これはドイツとの違いであり、打破しなければならないと感じている。

澤田 東京と横浜でもこれに関連した話題が出た。全国に展開し始めた団地再生については、国土交通省がコーディネーター(プランナー)補助金を出すようになった。日本では新しい政策である。一方のドイツでは、住民参加型ボトムアップ方式で、新エネルギーシステム導入の地域づくり等も始まっている。まちづくりの具体事例を比較し学び合うことが重要だと思う。

シュトレーブ 住民の参画は大変重要である。人間はプロジェクトへの参画を通じて、初めて地域や課題にかかわる自分自身との関連性を理解し、確認できるからだ。

参加者 ステークホルダーには利害関係がある。日本では行政主導で早く計画を進めることに重きが置かれる。ドイツでは、どのように利害関係を乗り越えていくのか?

仁連 出発点において、多様なステイクホルダーが利害関係を超えて

■ Discussion

NIREN Mr. Straeb, could you first explain the process behind how you examined and developed the urban planning method for the IBA project? Could you also give any suggestions regarding how you managed to organize various stakeholders during the examination and plan development process?

STRAEB It all starts with the common understanding regarding the population decline issue, it is most important to discuss the possible positive and negative effects that the plan and program may cause among the stakeholders at a state/country level. This will lead to a common understanding between all stakeholders that collaboration and corporation will be beneficial for every stakeholder.

NIREN First there is a step in which involved stakeholders sharing their issues. Unfortunately in Japan this process is not formulated within the planning process and many issues are presented directly by governments or municipalities. I believe this may be the difference between the German and Japanese approach.

AUDIENCE Anti-nuclear power plant movements have made progress in Germany, however in Japan these movements show.
　We understand that stakeholders from diverse backgrounds should collaborate with recognition of each stakeholder's role and function. However in local community meetings in Japan, this process has shown little progress. What do you think the differences are in comparison to Germany?

STRAEB Society is composed of individuals with different roles. Machizukuri activities concern not only political issues but each meeting acts as an opportunity to distinguish what activities each stakeholder can do .It is very important for each stakeholder to learn something from each other.

MATSUOKA Local municipal committee members have a common problem of being restrained by community-specific human relationships. This influences their decisions and it is a clear distinction between the two systems. The individuality of these groups should be developed as it is so in Germany for a more diverse community.

SAWADA Similar issues were pointed out at the Tokyo and Yokohama conferences. New government measures which subsidize the planners have been recently introduced to encourage the Machidukuri process. On the other hand in Germany, bottom-up approach town renovation projects which use new energy concepts have just started an innovative machizukuri system. There are many things that can be learnt from comparing the past Machidukuri projects of each country.

STRAEB Citizen participation is very important. Citizens are capable of understanding the relations between themselves and planning issues of regional areas through the participation of planning process.

AUDIENCE There is a relationship based on interests among the stakeholders. In Japan, local governments will lead projects in order to prioritize the speed in which the projects proceed. How do you overcome the different interests in Germany?

仁連 出発点において、多様なステイクホルダーが利害関係を超えて議論ができる課題を設定することが重要ではないか。

参加者 「サステナブル社会」という定義について聞きたい。国、地方、都市、地域というレベルによって、まちづくりの方法論は異なると思う。滋賀県は、古いコミュニティの存在が大きい。ドイツでのコミュニティの形態はどのようなものか？

シュトレープ ドイツにも様々なコミュニティがある。人口減少等が原因で統廃合されるコミュニティもある。かつて、21のコミュニティを1つにまとめるプロジェクトがあったが、課題ではなく自身のコミュニティの長所を共有することでうまく行った。

仁連 ドイツの経験に学びながら、可能な市民参加の方法をつくり出していく必要がある。本会で話題に出た示唆に富む事例から学び、実践的かつ学術的に進めていきたいと考えている。

issues of regional areas through the participation of planning process.

AUDIENCE There is a relationship based on interests among the stakeholders. In Japan, local governments will lead projects in order to prioritize the speed in which the projects proceed. How do you overcome the different interests in Germany?

NIREN As a starting point, I believe that it is important to set tasks in which diverse stakeholders can overcome their interests for the purpose of the communal goal.

AUDIENCE I would like to know more about the definition of "Sustainable Society". On each level, such as country, region, city etc., the approach of machidukuri would be different. In the Shiga prefecture, the presence of the older communities is large. What form do German communities take?

STRAEB Germany also has a diverse range of community types. During the period after the reunification, there were a number of communities which merged due to population decline. In the past there was a project which merged 21 communities into one. However that project succeeded due to the contribution of strengths rather than the sharing of tasks.

NIREN We have learnt a lot from the projects in Germany but we need to continue to exchange machizukuri methodologies and practices, so that we may develop an effective approach ourselves one day.

第5章　グローバル化するサステナブルなまちづくりの課題
Chapter5　Sustainable Machidukuri in Globalized Societies

澤田誠二
SAWADA, Seiji

日本のプロジェクト

首都圏 / Metropolitan Tokyo
(36,900,000人 / 13,555 km²)
- 神奈川県 / Kanagawa Pref.
(9,072,000人 / 2,416 km²)
- 横浜市 / City of Yokohama
(3,700,000人 / 727 km²)
- 洋光台 / Yokodai-Danchi
(4,373戸・DU / 209 ha)
- 汐見台 / Shiomidai-Danchi
(5,516戸・DU / 75 ha)

関西圏 / Kansai Region
(22,760,000人 / 27,340 km²)
- 滋賀県 / Shiga Pref.
(1,390,000人 / 4,017 km²)
- 観月橋 / Kangetsukyo-Danchi
(540戸・DU / 3 ha)
- 富山市 / City of Toyama
(420,000人 / 1,242 km²)
- 黒部市 / City of Kurobe
(42,000人 / 426km²)

ドイツのプロジェクト

- ノルトラインヴェストファレン州
State Nordrhein-Westfalen
(1,784,000人 / 34,080 km²)

- Aエムシャーパーク / EmscherPark
- ボトロップ市 / City of Bottrop
(120,000人 / 140 km²)

(旧東ドイツ地域 / Eastern states)
- チューリンゲン州 / Sate Thuringia
(2,330,000人 / 16,172 km²)

- ゲーラ市 / City of Gera
(100,000人 / 152 km²)

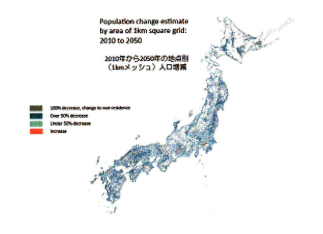

5-1 グローバル化するサステナブルなまちづくりの課題
Sustainable Machidukuri in Globalized Societies

澤田誠二
SAWADA, Seiji

はじめに

東京会議は"サステナブル社会のまちづくり"の本質についてグローバルな観点から議論する場であり、ドイツと日本の12事例が報告された。内容の豊富さと聴衆の多彩さはたしかにシンポジウムの目的に適っていた。しかし話題があまりにも多岐にわたり、参加者の多くがドイツと日本の状況の相対化に慣れていなかったためか、議論に必要な共通理解は十分ではなかったようだ。

これに対して横浜会議では、首都圏の神奈川エリアにおけるマスハウジングの歴史についての紹介と汐見台団地の視察とがあった。東京では多種多様な場所における事例が紹介されたのに対して、横浜では、神奈川エリアに場所を絞ってハウジングの供給や都市開発にかかわる諸問題の総括が行われた。

滋賀会議は東京、横浜とは雰囲気が幾分異なっていた。滋賀は琵琶湖を囲む山々の裾野にあり、環境科学で知られた大学が開催主体である。基調講演「滋賀のサステナブルなまちづくり」は、環境経済学の観点から"人口減期"における"自然環境資産、人工環境資産、遊休環境資産"の扱いについて提案する内容であり、参加者の多くは自治体関係者であった。

今回の国際シンポジュウムは、東京会議にこうした横浜と滋賀の会議が加えられた結果、"サステナブル社会のまちづくり"に関する議論は充実したものになった。

01 総論―ドイツと日本におけるまちづくり発展の状況

3カ所での会議におけるプレゼンテーションと討議の内容とは、次のようにまとめられる。

（サステナビリティ）

サステナビリティは日本の都市計画法には含まれてなく、自然環境、エネルギー、安全などのテーマ別に各省庁が管理することになっている。一方ドイツでは、プランナー（コーディネーター）がプランニングの要（かなめ）にあって、自治体・州・連邦レベルでのプランニングの調整に一貫して当たる。サステナビリティ（持続可能性）という複合的な計画目標の達成のためのさまざまなプランニングを統合・調和する役割を果たしている。

このサステナビリティについては、東京会議で下記のモデルが紹介された。エコロジー、エコノミー、社会生活の3分野の持続可能性を向上する努力がサステナビリティの達成につながるという。

このコンセプトを、シンポジュウムで紹介された事例を比較・評価する際の基準とした。

サステナビリティーを支える三本の柱

エコロジー
エコノミー
社会生活

（まちづくりの体制）

日独の事例は、国のレベルの大きなプランから比較的小さな、しかし

Preface

The Tokyo conference was an occasion to review presentations on projects and policy measures, and to discuss essential issues about sustainable Machidukuri. Twelve very well prepared lectures were given to the audience from different sectors. Thus as the contents were too much and widely ranged for the audience, and therefore meaning of international comparison was not be well understood by the audience. The discussion session compensated for it.

The Yokohama program, which consisted of the site visit to the Shiomidai housing project in the morning, and lectures in the afternoon on development history of housing supply corporation since the post–WWI period until today. Talks about actual works of planning, developing, and management of a number of housing estates in Kanagawa gave the audience a good opportunity to learn about the development of living, industry & economic situation and housing supply particularly in Yokohama area. Thus the participants could understand the complex problems of the symposium more realistically.

In contrast to the cityscape of Tokyo and Yokohama, that of Shiga is closer to real nature, Biwa lake. The keynote address "Sustainable Machidukuri of Shiga," was given by an environmental-economist's point of view, focusing on community development approach which deasl with natural environmental capitals, artificial environmental capitals and idle capitals. This lecture reflected well Shiga's position in today's changing Japanese society.

The aim of symposium was thus at the conferences in Tokyo, Yokohama and Shiga realized at last on 3. December 2014.

01. General Remarks – Development of Machidukuri in Germany and Japan in Comparison

The presentations and discussions at the conferences in Tokyo, Yokohama and Shiga can be summarized as follows.

(Sustainability)

In Japan sustainability is not a subject of town planning policy. Ministry of Construction, called today "Ministry of Land, Infrastructure and Truism, MLIT), controls town planning. The sustainability-related issues, such as energy, climate, nature and industry etc., are the subjects of other Ministries in Japan. German planners have to deal with sustainability issues, on project by project base, and to assist decision making of the municipalities. Sometimes the states would help that decision through specific subsidiary for realization of the project.

At the symposium, "sustainability" was defined that three pillars, i.e. "Ecology", "Economy", and "Social Life", are supporting achievement of "Sustainability".

Three Pillars of Sustainability

Ecology
Economy
Social Life

(Social systems of machidukuri sistem)

Japanese and German planners play proper function and roles

尖鋭なアプローチのものまで多様である。それらについて一定の観点から評価するには、国の運営システムの違いについて、あらためて知っておく必要もある。

国の運営体制

（日本）	（ドイツ）
政府	連邦政府
地方自治体（都道府県）	州政府(States/Laender)
市町村	地方自治体(City/Stadt)

（まちづくりの発展状況、1980～2010）

日本での国によるサステナビリティ達成の取組みは、住宅・建物レベル省エネルギー法（1980年）にかのぼるという。またこの時期には人口動態のモニタリングも始まっているので、サステナビリティが重視される時代のことと理解できる。その後の国の取組みは、対象を住宅・建物レベルから面的なまちづくりへと拡大し、2006年には「まちづくり三法」の改正があり、商店街の空洞化の抑制、ゾーニング規制の強化・支援による中心市街地活性化の推進の段階に入った。

この間ドイツでは"エコロジー、エコノミー、社会生活"へ取組みがスムーズに進んだわけではない。1980年代に始まった"エコロジー"への対応では、まず生活者の意識改革が必要になった。それがなければまちづくりが始まらないからだ。次の"エコノミー"への取組みでは、1990年の東西ドイツの統一が影響しプロジェクト資金が集まり難い状況になったし、"社会生活"面の取組みでは、やむをえず1980年代からの"クオリティ・オブ・ライフ(QoL)"をモットーにエコライフづくりを進めてきた。東西統合が連邦・EUレベルの社会変革につながり、それが地域コミュニティの急速な変貌を促したのである。

日本でサステナビリティが本格的な話題になったのは1990年代からだ。建築関連三団体（建築学会、建築士会、建築家協会）は環境建築憲章を制定し（2000年）、2011年にはサステナビリティをテーマにした国際建築連盟東京会議が開催されている。

しかし2011年に発生した東日本大震災後のエネルギー変革策と住宅復興の提案を参照すれば明らかなように、まちづくりシステムの包括的イノベーションが始まるまでには時間がかかった。そのためもあって現在の日本では個人・コミュニティ・まち・国の各レベルで同時かつ並行して取組む状況にある。

これからのサステナブルなまちづくりのイノベーションを比較するには、21世紀に入って顕在化している"エネルギー変革とインターネット社会化"というメガトレンドについても考えなければならない。すなわち、ドイツにしても日本にしても、1980年ころから、サステナブルなまちづくりのイノベーションを、それぞれの社会運営の中で進めて来たが、そのイノベーション活動自体がこのメガトレンドに影響されることになっているのである。

また、21世紀に入ってからのまちづくりでは、気候変動と公共支出能力の弱化とがイノベーション推進にとっての"壁"となっていることを考えなければならない。

		（三本の柱）
1980年代～	環境の時代	エコロジー
1990年代～	成長しない時代	エコノミー
1990年代～	グローバル化時代	社会生活
2010年代～	エネルギー、ICT	3本の柱が一体化

02 ドイツと日本を代表するまちづくりプロジェクトの特色

シンポジウムで紹介された事例の中からまちづくりプロジェクトを選び、その概要、アプローチ、成果を要約する。

in planning and realizing Machidukuri process. Their function and role are not the same but relatively different due to federal and centralized government, and multilayered local systems. Role and function of German planner are definitive as he works for "City / Stadt (Gemeinde)" which has power highest right and duty in machizukuri.

Japan	Germany
The government	Federal / Bund
Local governments-Prefecture	State / Laender
Cities Towns, Villages	City / Stadt

(Development of Machidukuri, 1980~2010)

The first step toward sustainability in Japan was done by the Act on Rational Use of Energy, 1980. At that time, a population monitoring system for urbanized and local regions was introduced. In the '90s, this was up-graded to a system of 1 km mesh, so that town planning could accurately adapt to the changes in local conditions. Soon after these policy measures for sustainable machidukuri in regional level," Three Acts for Machidukuri in Urban Area" were adopted, namely "Regulation of Large-Scale Stores", "Downtown Revitalization, and "Town Planning Act" all together in '90s.

As for the efforts to deal with the three pillars of sustainability, in Germany, a progress from energy issue to further steps was made during 80s, but that was actually not problem less . Because ecology-oriented lifestyle development needed first of all people's awareness about it. This stimulated people's attempts at "economic growth" caused a negative impression more significantly during the reconstruction time after the reunification in 1990. Regarding the "social life pillar," the "QoL (Quality of Life)" was already achieved in the '90s in Germany. Around the turn of the century, the nation began to explore further meaning of "QoL". It is to say that Germany experienced radical change of social system due to the re-unification of Western and Eastern Germany and because of EU politics development during 2000s–2010s.

In Japan, the period of 1990 to 2000 was the time to start with "Reuse of Built-Environment Stock." This means the change from long-lasting "scrap & build" scheme to "conservation and reuse" scheme. Three top associations of researcher, planner, designer, and construction industries, in collaboration, published "Environmental Building Charter" in 2000. In 2011, they organized UIA (International Union of Architects) Tokyo congress which was the first in Asia. That was at the same the beginning of efforts toward sustainable machidukuri in Tohoku.

disaster areas. These areas demand machidukuri of "three pillars" in proper combination and shorter time schedule.

We have to assess the development of Machidukuri efforts, from 1990-2010 in Germany and Japan, taking into account the mega trends; population fluctuation, energy reform and ITC revolution. Climate change and public finance shortage became also pressing issues since last ten years.

In these circumstances, Machidukuri-involved professionals of both countries look for new and effective system toward sustainable societies.

	Era	Three Pillars
980s~	Era of Environment	＊ Ecology
990s~	Era of No growth	＊ Economy
1990s~	Era of Globalization	＊ Social Life
2010s	Mega Trends: Energy, ICT Technology etc.	

02. Machidukuri Projects in Germany and Japan – their Outline, Approach and Achievement

17 cases i. e. Machidukuri projects and innovation measures were introduced at the symposium. 9 cases represent present situation of Machidukuri in Germany and Japan.

01：イノベーションシティ・ボトロップのまちづくり

　ドイツ西部ノルトラインウエストファレン（NRW）州のボトロップ市（人口：12万人、面積：100km^2）。重厚長大産業で栄えたルール地方の炭鉱町だがIBAエムシャーパーク（1985年〜1995年）により回復し、現在州のイノベーション・シティ・ルール政策による脱産業化指向まちづくりを進めている。

　市全域の都市計画図による人工環境資産（建物・インフラ）状況の把握からパイロットエリアを設定。個別住民ヒヤリングによって建築リノベーションとエネルギー変革への対応を促す。これと並行して環境技術新産業の誘致も進め、今は市内他区域への拡大のためのまちづくりマニュアルも作成中。新時代のボトムアップ型アプローチであり、アーヘン工大、Wuppertal 研究所の支援を受け、EMSなど新技術の導入も盛んである。

02：IBAチューリンゲン 2013-2023地域再生プロジェクト

　旧東独チューリンゲン州（人口：230万、面積：16,000km^2）は広大な田園地帯。ドレスデン、ワイマール、イエーナ、エアフルト、アイゼナハなどの歴史都市が点在する。

　これを対象にIBA方式による"広域再生まちづくり"が中間報告2019年、最終報告2023年の予定で進められている。2014年秋に250件のアイデアが集まり、5つのテーマ領域に仕分けられて、具体化担当のワーキンググループの作業に入っている。
（5つのテーマ：7頁参照）

03：日本の団地再生・まちづくりの現状と課題

　老朽化住宅団地の再生は、サステナブルなまちづくりの一大対象分野。総数2,000万戸規模と想定される団地再生プロジェクトの内の200事例について、その事業主体組織、住環境再生の課題、プロジェクト・アプローチなどの条件を把握・整理したレポート。それらのプロジェクト創生活動を概観すると：
　　−居住環境の長寿命化
　　−多岐にわたる課題の集約的解決
　　−サステナブルな社会づくり
という3つの方向性の"流れ"と捉えられるとしている。

　また、団地再生プロジェクトニーズの顕在化が明確になってきたので効率的に取組むためのプロジェクト評価システムが必要だとしている。

04：まちづくり建設産業の転換
　−「箱」の産業から「場」の産業へ

　過去40年ほどの間に他国が経験したことがないほどの新築市場が継続し、工業化、都市化の技術開発が行われてきた。しかしこの5年ほど前からは新築市場は衰退し、あり余るほどの住宅がある。（空き家は820万戸）。こうした建築市場の一大変革に対応して"箱の産業"から"場の産業"に変換する産業化"リノベーション"が始まりつつある。この新産業化はICT社会化の元に実施するオンサイト・スクールによるもので、全国に伝搬しつつあり、まちづくりに発展する可能性がある。

05：ゲーラ2030−まちづくりの統合的アプローチ

　東西統合後25年経ったが、2007年には連邦復興資金（公共財政赤字解消策）も打切られたのでゲーラ市（人口：10万人、面積：152km^2）は"再生まちづくり"を始めている。2012年には市議会で"まちづくり公社"主導とする議決を得て、サステナブルなまちづくりを2030年目標達成の予定で進めている。その戦略は下記4テーマごとに編成した市民のワーキング活動にある。

　　a.未来先取りの新産業おこし、企業誘致
　　b.前向きに行動する市民、公正な市民行動

01：Innovation City Bottrop Machidukuri Project

　The city of Bottrop, with its population of 120,000 and area of 100 km^2, is a former coal mining base, which was revitalized during 1990s through IBA Emscher Park project of the state NRW. (Nortrhein-Westfalen). Innovation City Ruhr is a state's machidukuri policy measure which started about 10 years ago, and Bottrop was nominated as one of them.

　The approach of Bottrop project is a typical "bottom-up" system. The project planner starts first individual dialogue with the citizen to examine his building retrofit plan and introduction of new energy concept in the "pilot project area" (70,000 people i.e. 14,500 households).

　After the verification of feasibility of the renovation and energy reformation by the pilot project, the city will expand that innovative town planning further to other areas of the city.

02：IBA Thüringen 2013-2023 Regional Development Project

　In the state / Land of Thüringen, population 2,3 million and 16,000 km^2 area, a IBA project is now going on. The Land is a very large hilly terrain, has a number of historic cities such as Dresden, Weimar, Jena, Erfurt and Eisenach. They are located independently with appropriate distance from each other. It is a traditional landscape which was formed in medieval time.

　The project started in 2013, its "interim report" is set due in 2019, and the "final report" year is scheduled for 2023. In 2014, the proposal entry of 250 were submitted, and five working groups were organized to review the proposed plans.

　The IBA Thüringen project addresses not only municipalities, it also addresses individuals, societies, associations etc. and invites them to make project proposals.

　Proposals need not necessarily end up with construction works, all other kinds of projects that could provide solutions (IT-solutions, organizational solutions etc.) are welcome as well. The main fields of intervention the project aims at are:
- Livable quarters such as quarters providing high urbanity
- Sustainable villages such as "independent cycle" life
- Productive landscapes such as farming etc.
- Landscapes for experiences such as for learning from the nature

03：Recent Housing Refurbishments Projects in Japan
　– Need to develop project assessment system

　The very large housing stock of Japan which was supplied in 1950s and 1960s, about 20 million units, is recognized now as the field for housing refurbishment. About 10 years ago a number of citizens' and/or municipal initiatives started to work on creation of housing refurbishment projects and programing of the projects.

　The data of these projects were collected by the Housing Refurbishment Association (HRA) and classified to three project groups with emphasis on:
　a. longer life of the built-environment,
　b. resolution of several newly coming-up problems of living
　c. access to full-scale sustainability development.
Some studies on the assessment and evaluation of the projects have started already on the initiative of the HRA.

04：Machidukuri Industry's Change – From "Box"-making to "Place"-making Industry

　Japanese building & construction industry achieved world-top level growth already in '90s. Builders, construction companies as well as building product industries are now in a phase of structural change. Mainly due to no-growth economy since 90's and the change of building scheme from "scrap & build" to "renewal & renovation." This was caused by saturation of new construction market.

　As for Machizukuri in residential building market, where the

c.応用技術の重視、教育研修事業の充実
　　d.アーバンライフ感覚あふれる都市空間づくり

06：富山市－"環境未来都市・コンパクトシティ"

　富山市（人口：42万人、面積：1,242km²）は北陸の基幹都市。製薬業を中心に恵まれた発展を遂げてきたが、人口減少と高齢化の進展と市街地の外延部への拡大への対応など考えて"コンパクトなまちづくり"を2000年より進めている。

　施策の中心はLRTの導入と外延部居住者の呼び戻し、都市的施設の整備などを都市経営効率向上と併せて実施するものであり、2011年に「環境未来都市」にノミネートされ、OECD「コンパクトシティ報告書」にも掲載されて、環境と超高齢化分野で世界的な成功事例として高く評価されている。

07：パッシブタウン黒部モデル

　著名な世界企業YKKのパッシブデザインの社宅団地（250戸、2025年完成）である。黒部市は立山連峰の北に位置し、豊富な地下水に恵まれた特有な気候の場所。低密度・低品質の自然エネルギー、太陽・風・地中熱をアクティブに利用して"心地よさ"を実現するエコ建築として初めてのものである。グローバルトップ企業YKKの先端的ハウジング提案として黒部市など地域自治体の住宅づくりに間接的に貢献できる。

08：横浜会議－神奈川県住宅供給公社の戦略、
　　　汐見台団地プロジェクト

　首都圏におけるまちづくりの担い手の一つ神奈川県住宅供給公社（1950年発足、70団地を経営）の活動の歴史の紹介と代表事例・汐見台団地の視察が行われた。

　神奈川における戦後から現在までの生活、産業の発展とハウジング・システムの進化とが明確に関係づけて紹介された。

　横浜会議は、中央集権システムによる日本の近代化過程を要約しつつ、戦後のハウジング開発の実態を要約して紹介するものであった。サステナブルなまちづくりについてグローバルな視点から考える良い機会であった。

09：滋賀会議－滋賀のサステナブル社会とまちづくり

　滋賀県（1,390,000人／4,017km²）における地域共生リーダー（滋賀県立大学・地域共生センター）の開催した会議であり、自治体関係者の参加が多かった。

　基調講演は、環境経済学の観点からのサステナブル社会のまちづくりをどのように進めるべきかという提案であった。すなわち、自然環境資産、人工環境資産、および遊休環境資産（空き家、休耕田など）の現状を把握し、現在進行する人口減、高齢化などの枠組みの中でどのようにプランニングするかについてである。この切り口は、東京会議、横浜会議での切り口に比べて、コミュニティのあり方に密着しているので、熱心な議論が行われた。

03　ドイツと日本におけるまちづくりシステムのイノベーション

01：ドイツの"サステナビリティ賞"

　2008年に発足した"トップランナー"顕彰制度。経済活動、コミュニティ活動、R&D活動、建設活動、企業活動の各分野ごとにサステナビリティ向上を評価し、その推進主体に賞を与える。2011年に"まちづくり"活動分野が加えられ（サステナブル都市コンセプト賞）、自治体はその規模ごと（大・中・小）に応募し、サステナビリティ向上努力を競い合う。このサステナブル都市コンセプト賞には"ガバナンス・行政面で優れる"、"気候変動・資源循環面で優れる"の2つの特別賞が設定されている。募集、審査状況、結果発表のメディア掲載を即時的にしてアピール効果を高めて

"vacant-houses" are 8,2 million in total, a radical change from "Box"-making to "Place"-making is currently taking place.

This movement is initiated by architects and real estate business people to organize direct communication on site between the house owners and the people willing to live there.

05：Gera 2030 – Integrated Urban Development Approach

The city of Gera, with its population of 100,000 people and area of 152 km2, in the state of Thüringen, had enjoyed after the re-unification a federal government subsidy for refurbishing public facilities and housing and infrastructure maintenance.

In 2004 when this subsidy program was terminated, the city Gera began to plan Machidukuri which can be implemented more independently. After a large citizen's meeting took place to start the project in 2000, a decision of city council was made in 2012 made about establishment of citizen's initiatives for planning and realization of the Machidukuri project.

The citizen's initiative consists of four groups which are now working on under following subjects;
  a. future-compliant working circles and strong economy
  b. socially-fair city with an active citizenship
  c. applied science-focused education and job training
  d. livable city with an attractive urban center

06：Toyama Environment Model-City Project

Toyama is one of the primary industry cities at Japan Sea coast region. It's population is 420,000 area 1,242 km². The mayor Mr. Mori started machizukuri project in 2000 to change its system adaptable to population decline. The trigger was

introduction of a new LRT (Light Rail Transit) line, which could attract the inhabitants scattered across the suburban areas back to the central district. This "Compact City" measure was already executed.

City of Toyama has been put on the list of Compact Cities, together with City of Kita-Kyushu from Japan, in the OECD Report 2012. Toyama has been also nominated Toyama received nomination of "Environment Model-City" from lately, as the city made a progress toward sustainability.

07：Passive Town Kurobe Model
– A future-oriented industry leads Machidukuri

The project is an employee housing of the YKK, a world famous Japanese industry. An ecology-oriented housing design was developed by Prof. Kodama, who is also well known in several international groups of researchers.

YKK has committed to Machidukuri of Kurobe city very long since its foundation during the reconstruction era.

This housing is considered to transfer to the city when the total housing stock of the city reach an appropriate level, namely its ownership would change from private to public.

The houses and open space are designed so that the people can enjoy the nature which is very specific in this area Kurobe, such as rich underground water, abundant sunshine in winter and good wind in summer. Mr.Kodama hopes that the interaction of life of company employees and citizens would stimulate the eco-life style and this model would expand to other areas of the city.

08：Yokohama Conference – Kanagawa Housing Corp's Strategy and Shiomidai Housing Estate Project

The conference constituted of three part; a presentation on the history of Shiomidai housing estate development and visit to the housing, an introduction about Kanagawa Housing Supply Corp. (KHSC) from its beginning 40 years ago until today, anda

02：ドイツの国際建築展(IBA)

20世紀初頭に始まった"建築や都市のプランニングと建設の実験を一貫して進め、総合的な検証"をするシステムである。"万国博覧会"は近代社会づくりをグローバルに先導して来たシステムであるが、IBAはテーマを建築・都市分野に絞り、集合住宅のモダンデザイン、戦後の住宅団地デザイン、都市の再定義・再構築、負の遺産地域の再生など、時代の要請に応えるイノベーション活動を誘導して来たシステムである。

IBAの企画・実施はそれぞれ有期の専従体制を整え、アイデアの公募、審査の公開、長期使用による成果の検証というサイクルを実行する。そうしたシステムが100年以上継続しており、イノベーションシステムとしての効用は高く評価されている。

03："スマートシティ"の推進

「スマートシティ」は"エネルギーとICTを融合して「賢い都市」をつくること"であり、2000年代に入って世界規模で多数のプロジェクトが動き始めている。日本には「世界の未来像」をつくる街「柏の葉スマートシティ」(環境共生都市＋健康長寿都市＋新産業創生都市)をはじめとして、各地のプロジェクトの公表や実証プロジェクトの報道がある。

スマートシティ・プロジェクトは必ずしも"既存コミュニティの維持・発展"に対応しないが、大企業がグループを編成してフロンティア市場を形成する活動なので、サステナブルまちづくりに必要なリソースが生まれることが期待できる。

04："環境モデル都市"、"環境未来都市"選定制度

内閣府(地域活性化統合事務局)の進める環境モデル都市とは、温室効果ガス排出の大幅な削減に向けてチャレンジする都市・地域を指し、富山市を含む23都市が選定されている。

また、環境未来都市は、環境・超高齢化に対応した人間中心の新たな価値を創造する都市であり、11都市・地域が選ばれている。環境モデル都市の選定は下記基準で5段階評価する。

1：取組みの進捗度
2：CO_2削減・吸収量
3：地域活力の創出
4：地域のアイデア・市民力
5：取組みの普及・展開

05：団地再生モデル補助事業

老朽化住宅団地の再生促進のための国土交通省補助事業で2013年に創設された。人口減少、高齢化への対応や低炭素型コンパクトなまちづくりの目的に適う萌芽プロジェクトを募集し、採択されたプロジェクトに対して3年継続の補助を行う制度。

この制度の具体策として住宅団地型既存住宅流通促進モデル事業が開始され、プランニング(コーディネーター)業務費補助プロジェクト22件が全国で進行中である。

サステナブル社会のまちづくりのイノベーションの現場が全国に設定され、モデル事業の進行状況を評価する体制が動いている。

04 総括―「サステナブル社会のまちづくり」研究のまとめ
―グローバル化時代のサステナブルなまちづくり

1980年代からのドイツと日本におけるまちづくりとシステム・イノベーションをまとめてみると、そこからサステナブルなまちづくりの方向性が浮彫りになる。
(ドイツ)

presentation on town planning issue specifically in relation to current population dynamics. Kanagawa has been the most important area since Meiji era. The history about KHSC touched it and presented how industry bases and housings have been planned and developed in general and in detail. It was a good introduction to Japan-specific housing development; 1950-2010.

09 : Shiga Conference – Community Development for Sustainable Shiga

Prefecture Shiga, 4,017km^2 large area, has Japan largest Lake Biwa at the center. The keynote was about vitalization of people's life in relation to capital of natural environment, artificial i.e. build-environment and idle build-environment, under current condition of population moves.

This aspect was different from that at the conference in Tokyo and Yokohama, and could stimulate good discussion about sustainable Machidukuri.

03. Machidukuri System Innovation Measures in Germany and Japan

01 : German Sustainability Award (GSA)

This sustainability award system started in 2008. The system aims at to promotion of sustainability achievement in five fields i.e. economics, community, R&D, construction and enterprise management. A award system for sustainable Machidukuri effort was added in 2011 which municipalities are to be granted. For evaluation of the efforts following five criteria are applied;
- Climate and local resources,
- Mobility and infrastructure
- Economy and labor
- Education and integration of local community
- Quality of life and urbanity (attractive urban environment)

02 : International Building Exhibition (IBA)

IBA system was founded at the beginning of the 20th century to promote innovative of planning, design, construction and use of built-environment. Program of selection of project and project realization i.e. "Interim reporting-phase", and verification of the project in real use i.e. "Final reporting-phase" are prepared for.

Announcement ant publication about IBA projects are to be made in each phases always open to the relevant expert, public organizations and citizens.

Subject of IBA system made significant move from architecture, urban environment to regional space, in terms of planning, design and project management since 1970s.

03 : "Smart City" Developments

"Smart City" is a new concept of urban development which was proposed along with ICT-based progress of society since more than ten years. Smart city projects are currently making shape in prominent urban centers worldwide.

In Japan, several smart city projects are going on by group of land owner, development, construction and service industries. Marketing company and banks are important member of the consortium.

The targets of these machizukuri efforts are new development.

The efforts would make contribution to sustainable Machidukuri by development of technology resources.

04 : "Environment Model-City" and "Environment Future City"

The Cabinet Office of Japan nominated previously 23 municipalities as "Environment Model-City", and 11 municipalities as "Environment Future City".

The "Model-City" is such city which is making intensive efforts

1. ドイツは、近代都市計画によってホモジニアスに形成された都市・国土を対象としてまちづくりの取組みが進められて来た。
2. ドイツのまちづくりの取組みは、都市（ゲマインデ）をサポートするプランナーの体制を基本としている。
3. このまちづくりシステムは、1990年の東西統合と2000年のEU結成との影響を受け、より合理化することが求められている。
4. そうした中で、ドイツ伝統のIBAイノベーションシステムなどを組合わせたサステナブルなまちづくりが進められ、一定の実績を上げている。

（日本）
1. 日本は、ヘテロジニアスに形成された国土・都市を対象とするまちづくりが、中央から地方へ、そしてプロジェクト拠点へというコントロールによって進められて来た。
2. 1980年代になってからは、地方分権化と機能の専門化が図られたが、普及実態は中央との距離により差異が大きい。
3. 21世紀になってからの日本は、少子化・高齢化など日本特有の課題も加わって、まちづくりの課題はさらに多重化している。まちづくりの進め方、イノベーションの進め方について、国と各地の状況に対応する方針の設定と、集約的マネージメントが進められている。

両国の1980年代からのサステナブルなまちづくりの進捗を概観すると、ドイツは社会変革の影響をうけながらも多数の実績を挙げていて、日本よりもおよそ10年は先行していると言える。
ドイツのまちづくりシステムのイノベーションはプランニング・マネジメントの改善に重点が置かれていて組織体制はあまり変更していない。そのことがスムーズな移行につながっている。

05　提言 -1 日本のまちづくりのあり方について

以上に整理した日本、ドイツにおけるプロジェクトとイノベーション施策に関する知見の中から、まず日本の参考になるものを下記のように選んだ。

01 "サステナビリティ" 指標に関すること
02 まちづくりプロジェクトの取組み方に関すること
03 まちづくりシステムのイノベーションに関すること
04 団地再生モデル開発事業の展開に関すること
05 プロジェクトの企画・推進方式の改善に関すること
06 国の主導、産業界主導のイノベーションに関すること

01：サステナビリティに関すること
サステナブルなまちづくりを進めるには社会一般の理解が必要である。
ドイツ・サステナビリティ賞は日本での啓蒙と共通理解の形成の参考になる顕彰制度である。この制度は、経済、コミュニティ、研究開発、建設、企業活動ごとに精緻な基準によりトップランナーを選定し顕彰する方式である。
この制度は、現在我が国で各分野別に進められるサステナビリティ達成の努力について、めりはりを付け、統合化する機能が期待できる。
日本の状況に照らして、この制度での活動分野の設定、評価と審査の方法そして応募条件の把握を行い、日本での制度と運営のあり方を設定してみることが望ましい。この顕彰によるイノベーション誘導効果はこの検討の過程で明らかになると考えられる。

02：まちづくりプロジェクトの取組み
日本における対象エリアを、大都市エリア、その郊外エリア、地方都市エリアの3つに分けるとすると、NRW州のボトロップ・プロジェクトは旧工業地帯に位置する自治体であり、京浜や北九州エリアの大都市

of reducing CO_2 emissions. The "Future City" is making efforts of reorganizing urban living environment to up-grading QoL in the city, and its adaptation to rapidly aging of the local community in particular.

The evaluation criteria used in the nomination procedure are;
1. Project Progress
2. Amount of CO_2 reduction – emission and absorption
3. Creation of local vitality,
   – development of local industries
4. Local idea and citizens' power,
   such as NPO and NGO organizations
5. Potencial of transfer of the efforts to other cities

05: Housing Refurbishment Model Project

This policy measure of MLIT (Bureau of Housing) is to promote refurbishment and renovation of existing housing estates. It is programmed so that planning, design, construction and management process should be innovated.

This measure includes subsidy program of project planning and building renovation. The selected proposals have to achieve proper project level within three years.
Open reporting of on-going project takes place every year to optimize measure program.

04. Summary of "Machidukuri in a Sustainable Society" Study – Efforts toward Sustainable Machidukuri in Globalized Societies

Characteristics of Machidukuri development in the period from 1980s to 2010s are;

(Germany)
1. German efforts have been done for homogenous city-and landscape which had been developed under the control of national, state and city organizations and by planner profession.
2. Machidukuri Efforts are so-called "bottom-up" approach that are played by municipality, planner and citizens.
  German Machidukuri changed its system slightly due to the reunification and the development of EU since 1990s.
3. Optimizing planning of overall the country and the local areas toward sustainable society is still kept through innovation measures such as IBA etc.

(Japan)
1. Japanese efforts have been made for heterogeneously developed land-nd city-scape across the country.
2. Japanese efforts had been controlled through network fromcentral, prefectural government and the cities.
The de-centralization of governance and empowerment of planner profession which was introduced about 20 years ago should be reconsidered as the Machidukuri paradigms have changed recently.
3. Along with recent radical change of society, i.e. population aging and decline, new Machidukuri system-innovation measures have been ntroduced previously.
  Some measures aims at finding target and approach ofproject at the same time. This is a change of policy measure itself. The change also involves public, private role sharing.

It is to emphasis that organizational set-up and planner's role have been not much changed and sustainable Machidukuri is still in progress in Germany, where as in Japan a lot of efforts to improve and/or innovate Machidukuri project and its system.

と郊外の中間の立地だと考えてよい。

ボトロップの特色は、住民との直接の対話によってリノベーションとエネルギー変革の促進を図りへ、パイロットエリアから市域全体へのまちづくりに広げて行くアプローチにある。

これはボトムアップ型の方式であり、NRW州の地域振興策にサポートされ、地域の産業界、大学・研究所の支援もある。

シンポジウムで紹介された他のドイツ事例に比べると、すでに目に見える実績が出ていることもあるので、我が国にとって直接的な参考になると考えられる。

03：まちづくりシステムのイノベーションのあり方

我が国の地方都市エリアのサステナブルなまちづくりにとってはIBAチューリンゲン・プロジェクトが良い参考になると考えられる。このプロジェクトは旧東独チューリンゲン州の5都市が展開する広い地域を対象とし、次の5テーマに対応するアイデアを募集し、優れたものは実施してその有効性を検証するという実験過程を組み込んだプロジェクトである。

　　a. 都市間ネットワークのサステナブル化
　　b. エネルギー変革への対応
　　c. 人口の流動（移入、流出）への順応
　　d. 対象エリア内の循環システムの創出
　　e. 現行規制の改変

このプロジェクトは、現段階では目に見える成果は出ていない。ドイツ伝統の建築・まちづくりのイノベーションシステムIBA方式にならい2019年の中間報告、2023年の最終報告を予定して始まった。

昨年度は250件のアイデアが集まり、それらを上記のテーマごとに分類した段階にあるという。

このプロジェクトを、たとえば東北地方から適地を選んで対象とし、プロジェクト化の可能性について検討することはプランニング改善に役立つと考えられる。その検討の過程で、国と県および自治体の担い方と連携についても検証できる。

04：団地再生モデル開発事業の展開

この事業は、老朽化した住宅団地の再生というまちづくりの促進策であり、全国各地から選んだ萌芽的プロジェクトについて、そのプランニング業務を助成し、その実務展開の結果の報告を集めて評価する集積する仕組みである。

このプランニングの出発点は、「空き家」の解消に例を採れば、団地経営あるいは住宅遊休資産の経営に関する住民参加の意見の調整である。

ヨーロッパ特にドイツで1980年代以降現在までに実施された団地再生まちづくりは多数存在すること、その背景には1990年の"壁の崩壊"があり、「空き家」現象を団地経営上の課題として取り扱ったことは、当時各国に詳しく伝えられている。

現在日本で始まった「団地まちづくり」イノベーションを効率の良いものとするには、欧州やドイツの先行事例をあらためてデータベース化する必要がある。それは個々のプロジェクトごとに、団地の概要だけではなく再生計画の進め方および団地経営データのわかり易い体系化が必要である。

05：プロジェクトの企画・推進方式の改善

団地再生を例にとって日本におけるサステナブルなまちづくりのあり方を予測すると、その対象プロジェクトの多様さがまず目に入る。

プランナー（コーディネーター）は、プロジェクトの目標とプランニングの進め方について住民や管理者と意見交換し、再生プランにまとめなければならない。

検討が再生事業コストに進むと、しばしば分担や事業リスクに関する

05. Suggestions about the Efforts in Japanese Circumstances

In reference to the above described innovation issues, the following points will give good teaching for Japanese innovation measures and projects.
　a. Definition of Sustainability
　b. German projects of good reference to Japanese projects
　c. German innovation measure which could be introduced smoothly
　d. Good practices for housing refurbishment projects
　e. Improvements of planning and project management systems
　f. Issues regarding governmental measures and industrial nitiatives

01 : "Sustainability" – Reorganization of the Efforts In various Sectors

Good public understanding about sustainability is the precondition for planning and realization of sustainable cities. German sustainability award (GSA) is a good model for promotion of sustainability in Japan. GSA is programmed clearly for each target groups with proper evaluation criteria. To study this model in relation to present situation of "sustainability" in Japan, in general and detail should make that sector by sector differentiated effort more integrative.

GSA includes a specific program for promotion of sustainable machizukuri. Its target group is municipality.

02 : Machidukuri Project Efforts which are able to adopt to Japanese circumstances immediately

In reference to Japanese Machidukuri effort fields i.e. large urban areas, its suburban areas, cities in local regions, Bottrop case in NRW seems to be a good model for the efforts in large urban areas in Tokyo, Osaka, Kitakyushu and Nagoya regions.

The approach of Bottrop namely "bottom-up" system town planning through "renovation(retrofitting) of urban environment and adoption of new energy concept" is exactly what those cities in the Japanese regions look for, and local munisipalities are prepared to support such Machidukuri effort.

03 : Machidukuri System Innovations which are able to adopt to Japanese circumstance – IBA like system

As for sustainable Machidukuri development in Japanese "local regions", German IBA system has a good model effect. The state Thüringen is a low economic growth region in comparison with NRW, Baden-Württemberg and Bayern.

There should be regions in Japan which are in a situation comparable with Thüringen. The cities and towns in Tohoku and Shikoku regions should have applicability of IBA-like measure. Following criteria of Thüringen case can be used for a study:
　a. sustainable networking of the existing cities
　b. adoption to new energy concept
　c. adaptability of community to population moves
　d. creation of circulation systems in local areas
　e. challenges based on existing standards

04 : Housing Refurbishment Model Projects

This is a measure to promote refurbishment and renovation of deteriorated large housing estates across Japan. This has high possibility of moving Machidukuri from local optimum solutions to overall optimum solutions if its cooperation with relevant innovation efforts would be managed appropriately and effectively.

The administration of the measure program, together with achievement of individual project, and development of cooperate measures should progress side by side and smoothly.

論争に発展する。

こうしたプロジェクト形成状況を、ドイツと日本について比べるとかなりの違いがある。

その違いの背景には、建築再生の経験の短さ、共用部分の使用と管理の基になる共同生活意識の弱さがある。サステナブルなまちづくりのためのシステムづくりと担い手の育成では、こうした違いを踏まえた先行事例の参照が必要である。

また、そのシステムをどのエリアー大都市、その郊外、地方都市ーに適用するかによって、そのプロセスが違ってくる。

日本ではまちづくりのニーズが膨大であり、多様なので、以上の観点を踏まえたプロジェクトのモデル化ープロジェクト評価システムーが必要と考えられる。

06：国や産業界の主導するイノベーションの役割り

環境モデル都市と環境未来都市は、スマートシティ・プロジェクトと並ぶ"3本柱"統合的な取り組みである。

これらの取組みが相乗効果を生み、他のイノベーション活動にも刺激を与えられるような情報交流システムを用意すべきだ。

ドイツのIBAやイノベーション・シティ・ルールの促進策には、ブランド化と関連知識の体系化とデータベース化が組み込まれていて、それが有効に機能している。

06 提言-2 ドイツのまちづくりシステムのあり方について

ドイツにおけるまちづくりについて、そのプランニングに関しては特定の提言は見当たらない。しかし、まちづくりの実施段階すなわち建設の段階に関して、次のことが指摘できる。今後の建築生産システムは部品化の方向にさらに進化するはずなので、現在の建物構成材システムの円滑な改善の準備が必要である。ことにEU圏内の諸国においてまちづくりに携わる際には、このことが必要になる。

というのは、EU諸国間そしてEUとアフリカ、アジア間の建築やまちづくり分野で交流がいよいよ盛んになると予想され、そうした交流の基盤になる建設段階の技術コンセプトが部品化システムだと考えられるからである。

07「サステナブルなまちづくり」研究の今後の展開

サステナブルなまちづくりは先進諸国だけではなく、急速に都市化の進む国々にとっても重要な関心事である。この研究は、ドイツと日本を対比的に捉えて、まちづくりの経緯を振り返り、その今後のあり方を展望した。

ドイツが連邦制であり、日本は中央集権的という点で国の運営システムは対極にある。しかし同時に産業の発展による社会成長の面では共通している。

これらを踏まえてまちづくりの発展の経緯を見てみると、日本は"折重なる課題"に柔軟に対処してきたと言える。本研究の結果ににはそうした柔軟な対処を示唆するものがあると考える。

「サステナブルなまちづくり」に関する研究が、ドイツと日本の比較という枠組みを超えて、この課題を共有する諸国の研究者や実務者との交流に発展することを期待したい。

05 Improvements of Planning and Management system of Machidukuri Project

The important subjects of innovation of Machidukuri project are planning and management systems. Training system of planner professionals and database system for Machidukuri should be established soon in a smooth manner in collaboration of existing organizations. Editing capability is the key issue in planning the data base system since the system must serve for diverse projects in Japan.

06 Innovation Leadership of Government and Industries and their cooperation. "Environment Model-City" and "Environment Future City" are government's measures for sustainable Machidukuri in Japan.

"Smart City" projects are initiated by industry sector. Both of them are promoting sustainable Machidukuri with "three pillars" scenario, on individual project base. Cooperation of these initiatives should be made for machidukuri development.

06. Suggestions about the Efforts of Machidukuri involved people in German Circumstances

Regarding Machidukuri system itself, both to reach localized optimum solution over all optimum solution there is almost no suggestion to those who are involved in Machidukuri in Germany. A possible suggestion should be to building realization process in machizukuri. It is because building component system should be developed and become soon popular in Germany and the countries where German Machidukuri system will be transferred.

07. International Machidukuri Study in the Future

The concept of "Machidukuri in a Sustainable Society" study was on the whole established this time. Its verification should be made in extended circle and further study beyond German-Japanese comparison should be programed near future. The committee of "Machizukuri in a Sustainable Sciety" Study will take coordinator's role.

6-1 講師プロフィール
Speakers' profile

内田 祥哉 うちだ よしちか　UCHIDA, Yoshichika

東京大学名誉教授。1947年東京大学第一工学部建築学科卒業、工学博士。逓信省、電気通信省、日本電信電話公社。東京大学教授、明治大学教授、金沢美術工芸大学特任教授、工学院大学特任教授等歴任。日本建築学会元会長。日本学士院会員。

Professor emeritus Tokyo University. Ph.D. 1947 BA Univ. of Tokyo, Department of Architecture Faculty of Engineering. Ministry of Communication, Ministry of Telecommunications, President, Architectural Institute of Japan. Member of the Japan Academy.

大村 謙二郎 おおむら けんじろう　OMURA, Kenjiro

GK大村都市計画研究室主宰、筑波大学名誉教授。1971年東京大学工学部都市工学科卒業。（財）計量計画研究所研究員、建設省建築研究所室長歴任。都市計画、土地利用計画、市街地整備、日独の比較都市計画等専門。都市計画学会国際交流賞受賞。

1971, Univ. of Tokyo. Operating GK Omura Urban Planning Institute, Professor Emeritus, University of Tsukuba. President of Organization for Housing Warranty. Research projects: Urban and regional planning in the area of shrinking and aging society, planning control in unplanned mixed-use area.

奥茂 謙仁 おくも けんじ　OKUMO, Kenji

㈱市浦ハウジング&プランニング 取締役東京事務所副所長。1984年東京理科大修士課程修了。（株）市浦都市開発建築コンサルタント、福岡事務所にて団地計画、住宅設計に従事。東京事務所建築室長兼副所長。日本建築学会、JIA、（一社）団地再生支援協会プロジェクト部会副会長。

Deputy Director of ICHIURA Housing & Planning Associates Co., Ltd., Tokyo Office. 1984, M.A. Tokyo Univ. of Science.(Architecture Faculty of Engineering & Science). Housing Estate planning and Housing design at ICHIURA. Organization: Architectural Institute of Japan and many others.

井上 俊之 いのうえ としゆき　INOUE, Toshiyuki

1981年京都大学大学院建築学専攻修士課程修了。1981年建設省入省、長崎県土木部住宅課長、国土交通省住宅局市街地建築課市街地住宅整備室長、住宅局市街地建築課長、住宅局住宅総合整備課長、住宅局建築指導課長、大臣官房審議官（住宅局担当）を経て2009年住宅局長に就任。2015年（一社）ベターリビング理事長。

1981 MA, Architecture, at the University of Kyoto. After the works at several posts in Bureau of Housing Ministry of Construction, he took up the post of Directors of Bureau of Housing, Ministry of Land, Infrastructure, Transport, and Tourism (MLIT). Chief of Bureau of Housing 2009. President of Center for Better Living, November 2015.

ヘルマン・シュトレープ　STRAEB, Hermann

1972年シュトットガルト工大卒業（建築・都市計画専攻）。1977年 GRAS(Gruppe Architektur und Staedtebau)建築都市設計事務所設立。アルジェリア住宅省アドバイザー。建築設計、都市計画、アジアの高地集落調査に従事。1994年より旧東独ライネフェルデの団地再生まちづくりに参加。ワールド・ハビタット賞受賞。UIA都市計画省授賞。

1972, TU Stuttgart, Darmstadt.,Foundation of GRAS * Gruppe Architektur & Stadtplanung, Darmstadt. World Habitat Award for urban trans-formation project. Dipl. of architecture at Univ. of Stuttgart, Architect for conceptual design in Lenz Planen und Beraten, Stuttgart.

松村 秀一 まつむら しゅういち　MATSUMURA, Shuichi

東京大学大学院工学系研究科建築学専攻教授（建築構法、建築生産）。1980年東京大学大学院工学系研究科建築学専攻修了、工学博士。ローマ大学、大連理工大学等で客員教授歴任。都市住宅学会賞（著作）。主要著書「建築―新しい仕事のかたち 箱の産業から場の産業へ」、（一社）団地再生支援協会・副会長、（一社）HEAD研究会副理事長

Professor, Dept. of Architecture, 1980, Graduate School of Engineering, the Univ. of Tokyo.Ph.D. Univ. of Tokyo (Dept. of Arch.). Visiting Professor at Rome Univ., Nanjing Univ. Publication; Place-Making Indu

澤田 誠二 さわだ せいじ　SAWADA, Seiji

団地再生支援協会副会長、明治大学サステナブル建築研究所客員研究員。1966年東京大修士（建築生産論）。1969～1970年ドイツで建築デザインに従事。1974～75年フンボルト財団研究のためドイツ滞在。1982年清水建設勤務の後、2000年滋賀県大を経て、2002年～2011年明治大学教授。専門は構法計画。日本フンボルト協会評議員。「サステナブル社会のまちづくり」の編著者

Vice-Chairman of Association for Refurbishment of Housing Estates. Member of Research Unit Sustainable Building at Meiji University. Ph.D. 1966, MA,Univ. of Tokyo. Alexander von Humboldt Scholar at TU Dortmund, Shimizu Corporation, Prof. of Univ. of Shiga Prefecture and Meiji Univercity.

ゲオルグ K. ロエル　LÖER, Georg K.

株式会社 NRWジャパン（ドイツ ノルトライン・ヴェストファーレン州経済振興公社日本法人）代表。チューリンゲン大学、国際基督教大学、東京大学、1987年ベルリン自由大学修士。コーポレートバンカー（ドイツ、日本、中国他）、日独事業提携プロジェクトのコンサルタント。重点：再生可能エネルギー、ロボティクス、医療技術。独・日・英語に堪能。

Head of the Japan office of the Economic Development Agency of the German federal state of North Rhine, Westphalia, NRW Japan K.K. ICU and Tokyo Univ. and M.A. FU Berlin in History and Japanese Studies. Corporate banker in Germany, Japan and China. Consultant on German-Japanese projects.(Focus:renewable energies, robotics) .

ラモン・ミラー　MILLER, Ramon

ゲーラ市副市長（都市開発・環境担当）。1991年ハノーバー・ライプニッツ大卒（都市計画、ランドスケープアーキテクチャー）。コンサルタント事務所（プロジェクトマネジメント）、1996年ゲーラ交通局経営管理室長、副市長。ドイツ都市連盟都市開発委員会委員。

Deputy Mayor for Construction and Environment, City of Gera. 1991. Graduated from Leibniz University of Hanover, urban planning and landscape architecture. 1996 Head of the corporate planning unit of the public transportation corp. Gera. Member of German Union of Municipalities, Urban development Committee.

中村 圭勇　なかむら けいゆう
NAKAMURA, Keiyu

富山市環境部環境政策課環境未来都市推進係主査。2001年富山市役所財務部資産税課、企画管理部情報統計課を経て、2008〜10年三井物産（株）に出向、穀物ビジネスに従事。富山市役所農林水産部農政企画課。内閣官房主導の国家プロジェクト「環境未来都市」構想推進担当。「プラチナ構想スクール」第1期生。

Senior Staff of Future City Initiative Section, Environmental Policy Division, Environmental Department, Toyama City Office. Financial Affairs Dept., Information and Statistics Div., Planning and Administration Dept, Toyama City Office. Experienced at Mitsui & Co., LTD(as an assigned employee).

小玉 祐一郎　こだま ゆういちろう
KODAMA, Yuichiro

建築家。神戸芸術工科大学教授。工学博士。1969年東京工業大学卒。同大学助手、建設省建築研究所・室長、部長を経て現職。エステック計画研究所主宰。建築や都市のパッシブデザイン、サステイナブルデザインの研究開発とその実践に従事。住まいの中の自然（丸善）ほか。建築学会作品選奨、グッドデザイン賞、JIA環境建築賞ほか。

Prof. Kobe Design University, Dr. Eng., Architect, JIA. A pioneer researcher and architect in the field of Passive and Low energy Design, awarded AIJ Architectural Design Commendation, Good Design Award of Japan and others. a member of PLEA, IEA . A director of ESTEC DESIGN Tokyo.

猪股 篤雄　いのまた あつお　INOMATA, Atsuo

神奈川県住宅供給公社理事長。（財）シニアライフ振興財団理事長。1975年国立シュトットガルト芸術大学修士（建築専攻）。黒川記章建築都市設計事務所、シティバンク、ドイツ銀行を経て山口篤雄オフィス設立、HSBC業務に従事。三洋マネージメント取締役CFO、明和地所（株）執行役員。

1975 Dipl. Arch. Staatl. Akademie der Bildenden Kuenste Stuttgart. After the works at K. Kirokawa Architects, City Bank and Deuesche Bank, INOMATA established his office for building development consulting in 1995. 1995-2004 project manager for HSBC head office building. 2012 chairman of Kanagawa Pref. Housing Supply Corp

筬 健夫　しとみ たけお　Shitomi, Takeo

神奈川県住宅供給公社専務理事。1976年早稲田大学理工学部建築学科卒業、1978年東京都立大学大学院修士課程終了、神奈川県入庁。総務部施設整備計画担当課長、平塚市都市づくり景観担当部長、企画部土地水資源対策課長、2010年県土整備局建築住宅部長。

Senior Managing Director of Kanagawa Pref. Housing Supply Corp. 1976 Waseda Univ. Department of Architecture. 1978 MA Architecture ,at Tokyo Metropolitan Univ. After the works at several posts in Kanagawa Pref. 2010 Manager of Construction and Housing Department.

鈴木 伸治　すずき のぶはる
SUZUKI, Nobuharu

横浜市立大学まちづくりコース教授。博士（工学）。1992年京都大学建築学科卒業。1994年東京大学大学院修了。東京大学助手、関東学院大学助教授を経て現職。専門は都市デザイン、歴史的環境保全。著書に『創造性が都市を変える』（編著／2010学芸出版社）

Professor, Yokohama City University, Dr. of Engineering. 1992, BA, Kyoto University. 1994, MA, University of Tokyo. 1995-2000, Assistant Professor in Tokyo University, 2000-2006, Associate Professor in Kanto Gakuin University. 2006-Present, Yokohama City University. He specializes in urban design and conservation planning.

原 大祐　はら だいすけ　HARA, Daisuke

NPO法人西湘をあそぶ会代表理事、（有）湘南定置水産加工代表取締役、関内イノベーションイニシアティブ取締役。1978年生まれ、青山学院大学経済学部卒。神奈川県総合計画審議会特別委員。

The Executive Director of the NPO SEiSHO Asobu Association, the Representative Director and President of The Shonan Marine Products Corporation and is a Director of the Kannai Innovation Initiative co.ltd. Special Advisory Committee Member of the Commission Of Enquiry for the Comprehensive Planning of Kanagawa Prefecture. B.A. in Economics of Aoyama Gakuin Univ.

仁連 孝昭　にれん たかあき　NIREN, Takaaki

滋賀県立大学理事・副学長。1971年大阪市大卒、1981年京都大修士（環境経済学）。日本福祉大学、滋賀県立大学教授。サステナブルな社会について研究し、エコ村ネットワーキング理事長として、滋賀県でエコ村のデザインに関わる。Ask Nature Japan理事長。

Vice President, The University of Shiga Prefecture,1981, MA, The University of Kyoto. Director, Ask Nature Japan. Prof. Fukushi University and USP. Studying sustainable society and engaging design of the eco-village in Shiga.

松岡 拓公雄　まつおか たけお
MATSUOKA, Takeo

建築家。滋賀県立大学教授（1996年〜）。1978年東京芸術大学大学院芸術学修士。丹下健三都市建築設計研究所等経て、2006年アーキテクトシップ設立。田園・ランドスケープ等研究、フィールドワーク活動を展開中。作品:札幌モエレ沼公園、ガラスのピラミッド、ソニー MEなど多数。建築学会賞業績賞、土木学会デザイン賞最優秀賞など受賞多数。

Architect, Professor at University of Shiga Prefecture. 1978 MA, Tokyo Univ of Fine Arts, Department of Architecture. Kenzo Tange and URTEC, Professor at University of Shiga Prefecture. Awards: Japan Creation Awards, Good Design Award(Grand Prize Moerenuma Park) and many others.

大月 敏雄　おおつき としお　OTSUKI, Toshio

東京大学大学院教授。1967年福岡県生。1991年東京大学工学部建築学科卒業、1996年同博士課程単位取得退学。1997年横浜国立大学工学部建設学科助手、博士（工学）取得。1999年東京理科大学工学部建築学科専任講師。2008年東京大学大学院准教授を経て 2014年現職。専門分野：建築計画・住宅計画・住宅地計画・ハウジング。主な著書：「近居・近居・少子高齢社会の住まい・地域再生にどう活かすか」学芸出版社 2014年、「3・11後の建築と社会デザイン」平凡社新書2011年など多数。

1993. Graduate School of Engineering, the Univ. of Tokyo, 1997, Assistant, Yokohama National University, PhD the Univ.of Tokyo, 1999, Associate Professor, Tokyo University of Science, 2008, Associate Professor, the University of Tokyo, 2014, Professor, the University of Tokyo. Research fields: Planning of architectures and homes, housing settlement planning, Housing system design. Publication: House & Home in Population-Change Era and Regional Redevelopment, 2014, Architecture in Post 3.11. Society, 2011

6-2 国際シンポジュウムの計画
Preparation works for the symposium

澤田誠二
SAWADA, Seiji

■ **サステナブル社会のまちづくりシンポジュウムの計画**

　この「サステナブル社会のまちづくり」国際シンポジュウムは、2011年開催の明大「サステナブル社会のまちづくり」国際セミナーと2013年の秋開催のJGCCシンポジュウム「サステナブル社会の建築・まちづくり」を継承するものとして企画された。

　10年あまり前から進めていた団地再生に関する研究の方も、この間に社会的ニーズが顕在化して、プロジェクト評価に関する研究の段階に入った。それと並行して、全国に展開する萌芽的プロジェクトを対象にする国交省の補助事業も始まった。

　サステナブル社会のまちづくりをテーマにする国際シンポジュウムでは、こうした団地再生分野の状況を日本側のプレゼンテーションに反映させることができれば、JGCCシンポジュウムでまとめられた諸問題について、より包括的かつ具体的な議論ができるはずである。

■ **国際シンポジュウム計画の予備的研究**

　シンポジュウムの企画パートナーを予定していた H. シュトレーブ氏との打合せによって次の基本方針を設定した。

a. 「まちづくり」をコミュニティ・プランニング、住環境計画・デザイン、その構築と維持管理の各行為を合わせた一連のプロセス全体を差すこととする。
b. 「サステナブルなまちづくり」の主目的を"持続可能な生活・産業・文化活動のための環境"をつくることと設定する。
c. プロジェクトとイノベーション施策については、この条件にふさわしい候補事例を日独双方から提出し、シンポジュウムの実施条件に照らして選定する。
d. プロジェクト事例、イノベーション施策の事例、講師などのプログラム概要が絞り込まれた段階で、出来るだけ広い分野からの参加者を想定した会場の設定、資料の準備を進める。

■ **国際シンポジュウム実行委員会**

　シンポジュウムを主旨に沿うよう企画・実施するため次のメンバーからなる実行委員会を組織した。
－澤田誠二／元明治大学教授、建築生産論
－大村謙二郎／筑波大学名誉教授、都市計画学
－H. シュトレーブ／ GRAS（ドイツ）まちづくり専門家
－猪股篤雄／神奈川県住宅供給公社理事長
－仁連孝昭／滋賀県立大学・副学長、社会計画論

■ **アドバイザーの助言**

　シンポジュウムの企画内容の妥当性について次の専門家に助言を求めた。（英文JGCCシンポジュウム報告書を送付）
NJ. ハブラーケン／オープンビルディング
K. クンツマン／国土計画・地域計画
N. ウィルキンソン／ Open House International誌

NJ. ハブラーケンの意見
　JGCCシンポジュウム報告書から、サステナブル社会のまちづくりを包括的に設定していることが解った。日本とドイツの状況の対比から、今後

■ **How the Machidukuri symposium came up**

　This symposium is an inheritance of the international seminar at Meiji Univ. in 2011 and the JGCC symposium in 2013. These German-Japanese comparative studies were programs to explore the problems which seem to be beyond that of refurbishment of large housing complexes.

　Since more than a decade we had been working on research and development of re-planning, -design of deteriorated housing complexes, and also re-organization of housing management systems in Japan.

　In 2013, the government adopted a new housing policy for the housing refurbishment project. It was a good moment for us to be involved in this policy measure to put to the test what we have been worked on.

References:
Meiji University International Seminar, 2011, January
"Sustainable Machidukuri" Lecturer: H. STRAEB
Symposium, German Culture Center, Tokyo, 2013.10.31.
"Architecture and Cities in Sustainable Society"
organized by JGCC, Japan-Germany Cultural Community

　The theme of the symposium, which was programed to take place in December 2014, should be "Sustainable Machidukuri"
　The presentations about real projects and innovations in this theme field should be organized at the symposium so that the experiences from both countries will contribute to the wide audience.

■ **Preparation for the symposium**

　The preliminary discussion among the people involved in the preparation, four directions were decided:

a. Machidukuri means a process to plan, design and construction of "built environment", including operation and management of the community and living environment.
b. The main objective of sustainable Machidukuri is to realize environments for sustainable life, economy and culture.
c. Several number of project and innovative efforts of the Machidukuri should be presented and discussed in a dialogue manner. Those examples from Germany and Japan are to be selected when the theme the program of the symposium become more concrete.

■ **A project team was established for the preparatory works:**

のグローバル化時代のサステナブル社会のまちづくりのあり方を引き出そうという今回のシンポジュウムのコンセプトは妥当なものと考える。

　今まで日本の状況についての情報提供がわずかであったので、こうした国際シンポジュウムの開催は有意義なものになると考える。特に、日本が建設の産業化の面で先端的なことをアピールできると良いと思う。シンポジュウムの背景やプロジェクトの説明を整理して、リアリティのある議論となるようにしてほしい。

K. クンツマンの意見
　まちづくりに関する日独の対話は難しい。日独は、文化と政治の面で大きく異なり、都市のあり方も違っているからだ。技術先進国日本のスマートシティはサステナビリティに含むのか？エネルギー変革についてはどうなのか？
　しかし、サステナビリティについて住環境の耐用性に絞るならば、意義深い議論になると考える。（詳細は右欄のコメント（独文）を参照）

N. ウイルキンソンの意見
　Open House International 誌における議論の現状からみて興味深いシンポジュウムになると思う。結果を知らせて欲しい。
www.openhouse-int.com

■ドイツ側のプロジェクト、促進策に関する事前調査

　シンポジュウムの開催日程、会場、実施予算の概要について見通しが立った段階で、シュトレープ氏とドイツ側のプロジェクトと促進策の選定に関する事前調査を行った。

調査結果（シュトレープから）

　日本では、主に縦社会であることによりサステナブル社会への転換が困難だそうだ。

　ではヨーロッパではどうかと考えると、"サステナブル"がはやり言葉であるが、納得いくほどの実績はないように思う。

　そうしたことから"サステナブルな都市をつくるためにサステナブル社会が必要"という議論の方向を考えると"サステナブルな都市をつくる努力をすると、それがサステナブル社会づくりに貢献できるのか？"あるいはその逆になのかと考えざるをえない。

　自分自身のまちづくり専門家としての経験を踏まえると現実の回答はどちらでもないその中間に来るのではないかと思う。つまり、地政学的変化を伴う脱工業化（近代社会構築後）の現代社会では、新築ではなく既存建物や構造の質を問うわけだから、その低成長経済化では、まちづくり活動の進捗は時間がかかる。

　したがってシンポジュウムのプログラムは、シンポジュウム参加者がシンポジュウムの目的を明確に理解できるようアクセントを付ける必要がある。つまりどのセクターの関係者を議論に招くのか、そしてそのためにどのようなプロジェクト事例やイノベーションの施策を採りあげるかという点である。
　ドイツにおける現状を俯瞰すると、25年前に始まった"東西ドイツ統合"については、現在もその違いは依然として大きい。つまり、まちづくりの変革について共通に紹介をするのが難しい。というのも、東独エリアでは人口減の抑制が思うように進まず、個人もコミュニティも困難な財政状態にあるし、ハウジングの経営組織についても少数組織に集約する動きが依然として進んでいるのだ。

　こうした東独に特有なまちづくりの好例が"ゲーラ 2030―革新的まちづくり"プロジェクトであり、このプロジェクトでは市民がプランニングと実施のプロセスで中心的な役割を担う。この事例は、日本の地方都市でのまちづくりと比較する意味があるかもしれない。

　また、ドイツでは"ドイツ・サステナビリティ賞"という顕彰システムが

SAWADA, Seiji／Tokyo, Prof. Dr., Project Management
STRAEB, Hermann／Dresden, Town planner,
OMURA, Kenjiro／Tokyo, Prof. Dr., Town planning
INOMATA, Atsuo／Yokohama, Housing corporation
NIREN, Takaaki／Prof., Environmental economy

■ Comments of the advisors

The team developed a draft proposal of the symposium, and asked three researchers for advices on the proposal.

HABRAKEN, John／Apeldoorn, Prof. Open Building
KUNZMANN, Klaus／Berlin, Prof. Regional planning
WILKINSON, Nick／UK, Director,
Open House International Journal
www.openhouse-int.com

HABRAKEN's comments:

　It may be wise state clearly what the purpose of the coming meeting is. If the purpose is to inform a wider audience it is important to separate in the agenda the necessary general information from new research and proposals.
　I am impressed with the amount of information about particularly the situation in Japan that could be of interest to a wider international public which remains poorly informed. Particularly the international Open Building network could benefit.
　Perhaps a way could be found in some other context to inform the international community about the interesting developments in Japan relative to housing in general and the OB in particular, including, of course, the implementation of the Long Life Housing Act.

KUNZMANN's comment:

　Grundsätzlich ist es wohl immer problematisch Stadtentwicklung in Japan und Deutschland zu vergleichen, die kulturellen Unterschiede sind doch sehr groß und die politischen Rahmen Bedingungen auch.
　Stadtplaner können wenig dazu beitragen, da die Nachhaltigkeit sehr stark von gesetzlichen Umwelt- bestimmungen, von Gesetzen und Programmen zur
　Isolierung von Bauten abhängt, von Preisen für Energie und Mobilität, natürlich insgesamt immer vom Markt.
　Die Trennung von Funktionen in der Stadt, Wohnen, Arbeiten, Einkaufen und Unterhaltung) rückgängig zu machen und "urban villages" zu organisieren, ist sicher eine Utopie. Die Macht der Banken und des Kapitals , aber auch der großen Unternehmen ist zu groß.
　In Deutschland hat die grüne Partei viel erreicht, aber eben auch nicht alle Hindernisse für eine nachhaltige Entwicklung beseitigen können. Daher gibt es auch keine große Stadt besonders nachhaltig ist und plant. Kleinere Städte tun sich etwas leichter. Insgesamt gibt es in Deutschland viele kleine erfolgreiche Projekte, doch in der Regel ist die Rhetorik eindrucksvoller als die politische Wirklichkeit.
　Eine neue Dimension ist die neue Liebe von Städten für "smart cities", in denen neue Technologien als Mittel zur Reduzierung von

6年前から始まっている。これはトップランナーを顕彰するシステムであり、さまざまな分野毎に賞を授けるものだが、3年前から"サステナブル都市賞"が設けられた。

これは都市規模ごとに応募する仕組みでその特別賞は"統治（ガバナンス）と行政に優れるプロジェクト"と"気候変動と資源面の対応"に優れプロジェクトに与えられる。2013年度の"サステナブル都市賞"はイノベーション・シティ・ルールのボトロップ市が獲得しているので、シンポジュウムでこれを紹介するのは日本の先端的プロジェクトとの対比にもなるので良いと思うが日本の状況から見てどうか？

ちなみに、イノベイティブなプランニングの進め方の面では、自分達プランナーの立場から見ると、相手は政治家だからその立ち位置の変更によって努力が無駄になるケースも多い。

しかし一方で、政治家の意向次第ではちょっとした"試し（実験）"の余地が生まれる場合もある。

こうしたドイツにおけるプランニング業務の一般的な状況を概観すると、例外的にイノベーションを支援したのはIBA方式（国際建築展）だ。シンポジュウムではこの方式の近年の展開について紹介したいと思う。

自分としては、今始まったばかりだがIBAチューリンゲンをシンポジュウムでとりあげるのが良いと思う。というのは、このIBAチューリンゲンではサステナビリティを広い領域のことと定義し、それ自体の実験を計画しているからだ。たしかにこのプロジェクトの成果は2019年と2023年（中間評価と最終評価）に出す計画であり、ビジュアルなものを提示できるわけではない。しかし"サステナビリティを発明する"あるいは"サステナビリティに向けたイノベーション"のインストルメント（手段、方策）としては興味深い事例だと考えられるからだ。

■国際シンポジュウムの概要

以上のことを参考にして、3会議の実施条件を設定した：

a.ドイツ事例は、IBAチューリンゲン、ボトロップ、ゲーラ2030の3プロジェクトとする。
b.会場は、東京、横浜、滋賀の3カ所とする。（下記のとおり）
c.プログラムは、アドバイザーの意見を参考にし、企画目的に合う内容とし、提示内容に対応する配布資料を用意する。
d.用語は英語・日本語とし、進行は実行委員会メンバーの共同として、会議参加者間の共通理解を促すようにする。
e.各会議の告知は、各実施主体ごとに多分野の関係者に対して行う。

```
              名称        実施主体
12月1日（月）  東京会議    実行委員会
12月2日（火）  横浜会議    ＋猪股（県住宅公社）
12月3日（水）  滋賀会議    ＋仁連（滋賀県大）
```

nergieverbrauch gepriesen werden.

I suggest once more to narrow down the broad field, for example either to
i) sustainable building technologies
ii) sustainable urban design
iii) urban energy conserving policies,
or to any other field of comparison. This would raise the interest of potential target groups and facilitate the comparison of Japanese and German approaches.

STRAEB's comment:

The rich material on your conference in October 2013 reveals a strong criticism concerning Japanese society, who's vertical structure is supposed to be a main reason for missing sustainability. But I have doubts that European countries would be much better, sustainability is a very trendy issue but the results are still not so convincing...

The question is: do we need sustainable society to create sustainable cities or can the effort to create sustainable cities contribute to produce better (sustainable) societies?

The truth will be somewhere between these to positions. (What we have to integrate in our strategic approach ois the fact, that in our post-industrialized times with demographic change, the latitude for change is very small since we have rather to qualify existing buildings and structures than to invent new ones.)

Concerning our symposium, I am asking myself where to put the accents in order to make the subject more tangible for the participants (what exactly is the target group?). The choice of the examples is very essential.

The difference between Eastern and Western Germany still is very significant. The fact, that shrinking population and difficult financial situation of private and public households (as well as the concentration of housing estates in few hands) are more pronounced in eastern Germany impose different solutions and strategies.

A good example how to deal with these specific problems in eastern Germany is Gera 2030 project, where the citizens play a central role both in planning and implementation.

We planners have to admit, that we are quite good in analyzing the problems but less performant in finding adequate answers. And politicians are often following other, rather short-term objectives. And therefore, there is little space for experiments.

One exception is the IBA (International Building Expositions) measure, which are actually quite popular in Germany. It could be interesting to involve the IBA Thüringen project (actually starting) in our symposium, since it defines sustainability in a very large spectrum and sees itself as a field of experimentation. Results would not yet be visible (only by 2019 and 2023), but it is interesting as instrument to invent and test innovative solutions for sustainability.

Since 2 years, the German government attributes a national prize of sustainability in the urban context. There are prizes according to the size of the cities; in addition to this, there are 2 special prizes: "Governance and administration" and "Climate and resources". In 2013, a special award was given to "Innovation City Ruhr"). Bottrop project is a apart of Innovation City Ruhr, and probably interesting for Japanese participants.

■ Outline of Symposium – Thee Conferences

Taking into consideration above advices, we decide an execution plan of the symposium;

a. 3 German projects; IBA Thüringen, Gera 2030 and Bottrop
b. Venues; Tokyo, Yokohama and Shiga
c Program able to involve the advisers' suggestions Good information package to prepare for the attendants
d. Language; English & Japanese
   Moderation; Project Team members in collaboration to facilitate common understanding by the attendants
d.Annoucement; each organizer for each region

Venue	Organizer
2014 Dec. 1st Mo. Tokyo conference	Project Team
2014 Dec. 2nd Tu. Yokohama conference	Yokohama Team
2014 Dec. 3rd We. Shiga conference	Univ. Shiga Pre.

6-3 都市を耕すこと―持続可能な生活・産業・文化
Cultivating Built Environment

NJ. ハブラーケン

LANDSCAPE DESIGN, Vol XX, 2012年 特別寄稿 （訳: 澤田誠二）

■ ハウジング、都市環境の捉えかた

自分は1960年代から集合住宅の順応性すなわちアダプタビリティについて研究を続けて来た。主な理由は、私の考える"自然な関係"[*1]を現代のハウジングに再び導入しなければならないと考えたことにある。昔から長く続けて来た我々の居住と住宅とが一つになるハウジング・プロセスを考えたからだ。持続可能な都市環境（ビルトエンバイラメント）についても、その修復や維持が一体化するプロセスが常に問題になる。

自分がハウジングのデザインを始めた当時、都市環境を構成するビルトエンバイラメントが廃物になるなどとは全く考えていなかった。我々人類は数千年前から、高度に複合的なシェルターをつくり何世紀も"長く使い続ける"生活を営んで来た。だから都市環境を構成するファブリックが廃物になったり、再資源化することなど考えないで来たのである。

我々は歴史的都市の構造に着目して、都市環境を構成するファブリックの"有機体的クオリティ"について研究して来た。個々の建築の、ゆっくりだが継続するリニューアル、改善あるいは増築などを見ると、建物には独自の自己生成能力のあることがわかる。つまり住宅や建物はこのファブリックという有機体を構成する要素ないし"細胞"だと考えられる。これが生きているから、細胞を構成するモノの移動や取替えだけでファブリック全体の変容に順応できるのである。

現在様々なタイプの建築が存在するが、それは建築デザインの発明があったからではなく、大抵の場合こうした"個と全体"間の相互作用が盛んになったためと解釈できる。すなわち建築のタイプ化が進んだのは自然な成り行きであり、アーバンデザインとは、時代の証しとして建築を配置してファブリックを際立たせる行為だといえる。いわば造園家が樹木を選んで苗床に並べ花を咲かせるのに等しい作業なのだ。

現在我々の周辺に見る都市環境の様相は過去のものと極めて異なる。5000年も昔から続いた人類の居住環境づくりは、突然のように蒸発してしまって姿さえ見えない。しかし我々の努力は続けられたのだから、時代や社会によってその達成度は異なると考えるべきであろう。そう考えることで、これから何を為すべきかが見えてくるはずだ。

こうした仮説によって自分は生活細胞と有機体という捉え方による現状分析を進める。細胞の外周はモノでできていてそれを介して有機体の部分を構成する。有機体にはそれ自体の自律性がある。こうした状況はどのような都市環境にも認められる。したがって、都市環境のあり方を考え、発展の方向を予想するのは、そうした関係性の本質を見極めることに等しいはずだ。

そうした関係性をこの2つの概念で捉えれば、長い歳月をかけたハウジング行為、すなわち住宅を建設し都市環境を形成する活動を簡潔に記述できる。歴史を経て現代に至る過程で、どのように継承されて来たか、あるは中断してしまったかを理解できる。したがって、この研究方法によって、我々が現在の都市環境のサステナビリティー持続可能性ーについて、どのように考えるべきかが明確になるはずである。

■ 都市環境という有機体、そのサステナビリティ

たとえば米国で、ワシントンからボストンに向けて空路を行くと、600kmほどの間に、明るく照らされた道路を俯瞰できる。それらの明かりは夜通し絶えることなく、幹線道路は大西洋岸に沿って北に延び、アパラチア山脈へと分岐する様子がわかる。行き止まりになっているの

Metaphors to work with

From the sixties of last century onwards I have advocated adaptability in housing construction. The major reason for doing so was to re-introduce, in contemporary built environment, what I have called the natural relation[*1]; the age old settlement process where inhabitation and built form were one. Sustainability in built environment has everything to do with that restoration.

The idea that built environment can be wasteful is peculiar to our times. For millennia humanity has produced shelter that could be highly complex and endure for centuries. No one has ever argued that such fabrics wasted available resources.

The constant aspect of historic urban fabrics was their organic quality. The term is justified because by a slow but continuous process of renewal, improvement, and adaptation of individual houses, they had a self generating ability. Houses functioned like living cells of the fabric. There is a living cell where the nominal social unit interacts without mediation with the smallest material unit recognizable as a changeable whole. House types never were architectural inventions but came to full bloom by that very interaction. House typology being self evident, acts of urban design were about providing context for a bespoke fabric to manifest itself; much in the way a gardener lays out the beds for plants to flower in.

Our contemporary built environment may look very different from past examples, but it is unlikely that humanity s settlement habits of more than five thousand years have suddenly evaporated. It is reasonable to assume that they have continued and that the degree in which that has happened may tell us something about present day environment s health.

Following up on that assumption, I will look at present day built environment by using the metaphors of living cell and organism Living cell , as already stated, denoting those
instances where occupancy interacts immediately with its material envelope. Organism because a certain autonomy, a capacity for self organization, is found in all built environment, which we must accept as intrinsic to it. These terms provide a shorthand that can help us to understand age old human settlement behavior and consider its continuation and interruption in contemporary environment. This may help us to find out why we worry about sustainability in the first place.

Built environment as organism

A traveller flying at night from Washington DC to Boston Massachusetts, a distance of some 400 miles, sees a pattern of lighted roads and buildings that never darkens and which, following the Eastern seaboard, stretches inland to the Appalachian mountains.

There are spots where the network becomes vestigial while other places are a blaze of light. Already in 1961 geographer Jean Gottmann named it Megalopolis.[*2] From the air it appears as a giant weed on the surface of the earth with a life of its own, consuming energy and many other resources on a vast scale, day and night.

We know that today the world is replete with similar fields. The Mexican Federal District, the Nile delta triangle including Cairo and

は、かつての街が消滅し、別の場所に輝きが移ってしまったからだ。

　こうした様相について、1961年に地理学者のジーン・ゴットマンがメガロポリスと名付けた。空中から眺めるとまるで巨大なつる草が、特有の生命を持ち地表を覆っているように見える。そして、今日ではその生命体が多量のエネルギーと膨大な資源を日夜消費している。

　現在の世界は、こうしたメガロポリスで溢れている。メキシコの大都市圏がそうだし、ナイル川下流のカイロとアレキサンドリアを含むトライアングル地帯も同様だ。中国では香港の対岸の広東省にも同じ風景が見られるなど、その枚挙にいとまがない状況だ。

　それらはこうした複合体を形成し活力に満ちてはいるが、成長管理が行き届いてない状況にある。

■都市の"生活細胞"：その変容と消滅の系譜

　本稿でいう生活細胞とは、人間の居住活動が直接カタチに表れるもので、居住という行為は、自然資源の利用と自然生態系の律動性を考慮し決定する作業だと考える。インフラと呼ぶもの、特に交通・エネルギー・水・通信など多数のユーティリティ・ネットワークがこの居住活動をサポートする。そうしたインフラは生活細胞の技術進歩を許すものであり、インフラと細胞が一体となって"アーバンファブリック"という有機体を形成すると考えられる。

　歴史をさかのぼると、この"細胞"の基は"住宅"である。住宅は家長が支配し、場合によっては召使いや奴隷を加えた家族を包み込むものであった。当時は、労働のスペースも住宅の内部にあり、販売や取引き、あるいは食品加工や手工芸の場所もそこに含まれた。住宅は、いわばもっとも身近にある都市環境であり、一つの文化活動を代表する場所であった。

　こうした"細胞"という存在は、やがてベニスのゴシック様式の広場へ拡大したし、18世紀の北京には、ポンペイ風中庭のある細胞が細い街路によって階層化されてつながり、街区へと広がるファブリックが形成された。

　今日でも住宅は依然として生活細胞であり、郊外の都市拡張エリアには、富裕層のために近隣センターを含めた都市環境的空間がつくられて来た。それと同じ時代に、世界中のメガシティにおいてはそうした秩序感のあるものではないハウジングが形成されている。

　後者の事例を概観すると、こうした人間居住の衰退あるいは弱体化の様相が次々に表出している。そこでは生活細胞としての住宅の全てが場当たり的につくられる低成長期の産物だったり、"ファミリーの隠れ家"になってしまっていて、それに地域特産品の加工や貯蔵の場などが混在するという有様なのだ。

　こうした劣悪な都市空間を構成するのは、コンクリート、ガラス、レンガなどの建築材料、タイルや窓サッシ、ドアなどの建築部品であり、各種の配管・配線類であり、浴室・キッチンなのである。すなわち現代のアーバン・ファブリックの全てにおいて"飽食の時代"的状況が見られると言ってよいだろう。

　現代の住宅には核家族が住む。多様化の進む時代の社会システムの原単位である家族が住むのが一般的だ。その他の社会システムである仕事場、リテール・ショップ、セラピストアトリエ、コンサルタントオフィスなど多種多様なものがこの住宅と混成し、日々の交流と交際が行われている。

　すなわち歴史的に、多様化し緊密に形成されて来た社会的活動の"場"が"地域"を出て、さらに多様化し、増殖し続けて、現代ではアーバンファブリックに溶け込んでしまっている。"居住と細胞"という"モノ"の間には、より豊かな相互作用が期待できるとされていて、さらに多様化が進むようにも見える。

　かつてないほど小さくなり、しかしダイナミックな社会システム単位は極度に多様化し、今までの住宅中心とは対照的な発展を見せる場合も

Mexican Federal District, the Nile delta triangle including Cairo and Alexandria, the Guangdong area opposite Hong Kong, and many others come to mind. They are at least as complex and energetic and uncontrollable in their growth.

The living cell: change and disappearance

The living cell is where human settlement most immediately shapes physical form, and decides on the use of resources and the rhythm of environments metabolism. All infrastructures, particularly the many utility networks, serve to feed it. They also enable new innovations to reach that cell. Cells and infrastructure together make the organic whole.

In history, the house as living cell, under unified control, could encompass an extended family and might include servants and slaves. Spaces to work in were found in it as well:
To do business, to sell merchandise, to process food or do handicraft. Its contents represented a culture at its most intimate scale. As such the living cell could be large as the Venetian Gothic palace, extensive like the family compound in 18th century Beijing, hierarchically ordered like the Atrium house in Pompeii.

Today, the house is still a living cell and still determines built fields that range from suburban sprawl including neighborhoods for well-to-do citizens to the informal settlements surrounding most of the worlds mega cities. Of those, the latter most faithfully continue the age old process of human settlement where the house as living cell, for all its improvisational emergence and slow growth, harbors extended families and often combines domestic and work space. Informal settlement is now consuming materials like cement, glass, and bricks; elements like tiles, window frames and doors; and all kinds of piping and wiring, bathroom and kitchen equipment and much more - partaking of the same nourishment for all other contemporary fabric.

In today s environment on the other hand, the house, inhabited by the nuclear family, contains only one of many small social clusterings in need of a responsive envelope. Other social entities are found in the wide variety of work places, in retail stores, in the office of the therapist and consultant, the apartment building, and where else daily human intercourse takes place. The diverse and tightly packed social activities found in the historic house have escaped the domestic realm and, having diversified and multiplied, are found all across contemporary fabric: promising a richer and more diverse interaction between inhabitation and physical form.

In contrast to this rich diversity of ever smaller but dynamic social clusters, the smallest material unit recognizable as a changeable whole in the environmental fabric is no longer just the single house, but also the larger, sometimes very large, building. The result is an increasingly coarse fabric, unresponsive to inhabitation of any kind and at worst oppressively deterministic. The one-on-one interaction between material form and social entity, so naturally manifest in historic fabric is no longer evident. Many social clusters do not inhabit a form that lives by their care and initiative, and which in turn stimulates and facilitates their actions. The living-cell-that-could-be has been weakened or has entirely disappeared. Inhabitation for both work and living can now be found as a nomadic activity in a environment that is no longer its product or responsibility.

The fine grained large project

This lack of reciprocity between an increasingly diverse and dynamic social body and a less and less agile material environment is not just inconvenient. It complicates and slows down the metabolism that sustains built environment; and renders inhabitation passive and hence un-productive; leaving it incapable to stimulate innovation or

ある。すなわち、都市環境のなかで変化可能あるいは順応可能な最小単位は、一戸の住宅ではなく大きな建物つまり巨大なビルになるケースもある。

その結果アーバンファブリックは、次第に雑然とした粗末なものになって来た。それが結局居住活動を"無責任"なものにし、最悪の場合には"抑圧的"に都市環境が決まっているとも言える。

歴史的な"モノ"と社会システムの一対一の対応と相互作用という関係は形骸すら残らなくなった。いわゆるコミュニティと呼ぶ社会システム単位の多くが、今では主体的には運営管理されていない。住人をその活動を受入れなくなっている。つまり"生活細胞"は本来の機能を失い、その姿を薄め、場合によっては、完全に消滅することもある。

現代では、生活と労働そして居住の一体化した事例は、大自然の中に暮らす遊牧民(ノマド)が残るだけである。それらは現代社会が生み出すものでなく、またその社会的責任も担っていないのだ。

■小粒子の集積する都市開発プロジェクト

社会システムが多様化してダイナミックに発展する過程で機敏さが失われて来た現在、都市に集積したモノの相互間の互恵的な関係も次第に薄まって来ている。それは単に不便さを招くばかりではない。互恵関係が欠如すると都市環境の維持・保全という生態システムが複雑化し、後手後手に回る事態も発生する。つまり居住という行為は"受け身"になっていて、建設的ではなくなってしまう。そうした状況では技術革新も発生せず、住民やユーザーが刺激を受けることもない。すなわち"生活細胞"というものが、ファブリックを構成する諸要素の中で真の"自己再生"システムであることを再認識しなければならないのだ。

この生活細胞について言及するのは、今盛んな大規模プロジェクトを否定しようというわけではない。むしろそれらは現実世界の重要部分であり、我々はそこから逃れることはできないという認識なのだ。プロジェクトは、今後さらに大型化するはずだから、我々はそれを共有しなければならないと考えている。

こうしたプロジェクトで取組むべき課題は、生活細胞を包含でき、細胞の自己再生をゆるすようにデザインされた都市環境であろう。そのようにして都市のファブリックと生活細胞との自然な関係はよみがえり、小さな粒子の集積する"ファイングレイン集積型"の都市開発プロジェクトは成立することになるだろう。

今後は都市全体の複雑化がさらに進み、そうしたプロジェクトでは"生活細胞への回帰"の挑戦が併行して集約的に行えるはずだ、と私は考えている。

今日の生活細胞すなわち住宅のあり方を見ると、地域性は特に反映されず、浴室やキッチン設備、移設できる間仕切り壁、給排水の配管、電気・通信の配線、換気・暖房などで溢れている。こうした設備によって住宅と各室と仕事場は機能している。今日の居住単位は極めて多様化し分散化しているのだ。以前の緊密な関係でまとまった"世帯"に比べると小型の単位になっているが、人の生きる"シェルター"としては技術的にずっと複合的なものになって来ている。

こうしたファブリックと生活細胞の相互関係の消滅あるいは欠如の問題は、建築活動を担う者の職能の課題ととらえる必要があると私は考えている。住宅の品位の低下やあからさまな消失には、モダニスト達の指向性との深い関わりを私は感じている。すなわち、環境を構成する諸要素をただデザインの対象として捉えるだけで、規模の経済に依存して取組んで来たことを今や見直す必要があると考える。

そもそも住宅やハウジングは、デザインしたり、生産したりするものではない。専門家である我々が出来るのは、住民が自分で決められるように条件を整備することではないか?この意味で、現在の投資環境への取組みでは、"耕作"や"培養"といった戦略が必要であり、これは今までの生産システムとは異なるのだ。

to be stimulated by it.

Only a fabric of living cells can be a truly self generating system.

To be sure, pleading for the living cell is not a rejection of the large project, which is very much part of our world, is here to stay and most likely will be larger and more common.

The challenge is to make it accommodate the living cell, help the cell to take care of its own. Such a new way of restoring the natural relationship will produce what can be called the fine grained large project.

This challenge is intensified by the growing complexity of built form as such. Today's living cell, domestic or otherwise, must contain not only bathroom and kitchen equipment and replaceable partitioning, but also a host of conduits for sewerage, water, electric power, data, ventilation, heating, and more. These utilities are distributed throughout habitable space to serve every room or work station. Today's entities of inhabitation, so widely diverse and dispersed, may be small and singular compared to the closely packed historic household, but the envelope they want to animate is technically far more complex.

Before we look at this lack of reciprocity as another professional problem to be solved, we should pause and remember that the degrading or outright disappearance of the living cell had much to do with the Modernist inclination to see all things environmental as a technical problem or, at least, a design problem, and its believe in the economy of scale.

Living cells cannot be designed or produced, we can only provide conditions by which they may come about. To meet today's environmental challenge, a strategy of cultivation is required, not a production program.

The return of the living cell

If built environment is an organism, and as such has a certain autonomy, it can be expected to seek to remedy what is ailing it. And indeed, our present environment has been moving already in that direction almost unbeknown to its observers and theorists. A gradual introduction of the living cell in parts of our environment where it could not be found earlier on, can be noted:

For instance, the commercial office building, increasingly across the world, offers empty floors for the various occupant parties to have them fitted out by their own designers and fit-out contractors. There already, inside the large built volume, a clear separation of the collective environment on the one hand and the small scale act of inhabitation on the other hand, enables the living cell to return in a new way. In the same manner, an institutional client will expect a new office building to be able to respond to future mutations in work unit composition and location.

In the shopping mall, retail space is likewise left empty to be fitted out by specialized contractors familiar with the tenant retailer's house style.

So called open residential projects have been done on an experimental basis for quite some time now, [*3] but in the last decade or so, the approach has entered the commercial world. Residential Open Building projects initiated for profit can be found, among other countries, in Japan, Finland, the Netherlands and Switzerland. [*4]

When conditions are favorable the living cell appears spontaneously. In Moscow for instance, wealthy apartment owners would rip out the entire elaborately finished interior of their newly bought apartments to start again with their own designer. Appalled by this destruction of capital, developers, unaware of any international network advocating such an approach, now routinely produce "empty" apartment buildings, selling available floor space and collective facilities. [*5]

■都市の"生活細胞"の"回復"と"復権"が鍵

　都市環境が有機体であるなら、それ自体の自律性があるはずだ。それを理解すれば生活細胞を患わせているものをどう矯正すれば良いかわかる。実際に、現在の我々の周辺環境そのものが、観察者や研究者の気付かない内に、そうした方向へと動いているように思う。今まで気づかなかった都市活動細胞のゆっくりとした導入などは、そうした現をみることができる。

　たとえば、世界中に普及しているオフィスビルではテナント企業は自分の選ぶデザイナーと工事会社によってその内装部分を完成させる。つまり大きなビルでは共用スペースとは明確に仕切られた広い空間が、様々なテナントの活動単位ごとに区切られるのだ。こうした方式によって今までより活動細胞を活性化させることができる。

　役所関係のビルでも、同じ方式によって、将来の部署構成の変化や配置替えへの対応が出来るようになって来ている。

　ショッピングモールのショップ空間のあり方を見ても、テナント・リテールの小売店のスタイルに詳しい専門家がオープンスペースを埋めるようになって来ている。

　いわゆるオープンビルディングと呼ぶ住宅・ハウジング分野の研究開発を振返ると、30年ほど前から実験プロジェクトが相当数行われて来ている。その結果、この10年ほどの間に、実際のプロジェクトに展開されるようになっている。プロジェクトの多くはビジネスの効率向上を目指し、すでに様々な国で展開されている。中でも主要なプロジェクトが日本、フィンランド、オランダ、スイスで実施された。

　集合住宅の場合、プロジェクトの環境さえ整っていれば"生活細胞"は自然に生まれてくる。たとえばモスクワには、富裕層向けマンションの入念に作られたインテリアを、購入者が全面撤去して、自分の好みのデザイナーにあらためて作らせたことがあった。この"資本の解体"に驚いたデベロッパーは、このオープンビルディングの動向をそれまで知らなかったのだが、現在では、オープンスペースのアパートを建設し、床スペースと共用スペースを合わせて販売という方式に切り替えたという。

　病院建築では、病室や手術室側の要求で決まる部分が多く、同時に全体として柔軟な適応が常に求められる。スイスの病院の例では、まず医療専門家の議論があり、それを受けた様々な施設側の技術革新が実施される。中でも最も先端的なプロジェクトは首都・ベルンの集中治療病院の場合である。この建物では、いわゆるスケルトン部分とインフィル部分の徹底した分離が設定され、それぞれについてのデザインをコンペ方式で募集し、実施案を決定している。

　以上のような事例を概観すると前述の"小さな粒子の集積するきめの細かな大規模プロジェクト"に伴う新たな専門家の登場していることがわかる。

　これに関しては、内田祥哉が先鞭を付けた。すなわち内田は、オープンビルディングの先駆的プロジェクト・NEXT21においてデザインチームのリーダーを務めた。大阪に実現したこの集合住宅プロジェクトにおいて、彼はフレキシブル建築を計画しただけでなく"3次元ファブリック（立体街路）"を試み、同時に13名の建築デザイナーによるNEXT21の18世帯に対応するデザインのカスタマイズを誘導したのだ。内田自身は、個々の住戸を収容するインフラすなわちスケルトンのデザインとパブリックスペース部分のデザインとを担当した。

　同様な建築家の新しい役割が、フィンランドのオープンハウジング・プロジェクトにも見出せる。この事例では、建築家はデベロッパーをサポートし、入居希望者がインテリアのデザインと仕様を個別に注文できるようにし、全体の建設工事を管理している。

■都市化の戦略-1：長期的投資のプランニング

　このように今後の都市環境の発展について展望すると、次のような

In hospital construction, the perennial need for partial and ad-hoc adaptation of space in response to the changing demands of work units has triggered debate in professional circles and has led to various innovations. The most advanced project, so far, is found in a new intensive care hospital for the city of Bern where a radical separation of the so called primary system from the secondary system led to separate design competitions for each.[*6]

All this leads to an awakening awareness of a new professional role implied in the design for the fine grained large project. This was perhaps first expressed when Professor Yoshichika Uchida, the leader of the design team for the path breaking NEXT21 residential open building project in Osaka, Japan,[*7] declared that he did not want to do a flexible building but a three dimensional urban fabric. True to this statement he invited thirteen other architects to custom design the eighteen dwelling units in the project; occupying himself with the design of public spaces and infrastructure for the living cells to settle in.

A new professional attitude is also found in a Finnish company dedicated to help developers to serve would-be apartment buyers in the custom design and specification of the fit-out of their units, and to oversee actual implementation.[*8]

New opportunities: long term investment

Once this new perspective on urban fabric unfolds, it becomes possible to point out some opportunities worth exploration:

The base building, as it is already known in commercial development, now free from demands from individual inhabitation, can live a long time. Just as streets tend to live longer than houses. This allows for long term investment which, in turn, makes higher initial investment feasible for internal public spaces and external architecture. Right now this approach is followed by a not-for-profit housing corporation in Amsterdam where a urban block containing seven such base buildings, here called solids, each nine stories high, is under construction. Empty space will be for rent and occupants are free to take care of their interior fit-out for any purpose that is not disruptive to the community. The extra investment in high quality facades an exterior arcade and inside public spaces, is expected to bring a profit in the long run.[*9]

No doubt this initiative will be regarded with skepticism by many. However, it is in tune with a law, passed by the Japanese congress in the year 2008, which encourages residential construction that can last up to two centuries. The law refers to detailed technical requirements in which subsystems are identified that must be replaceable to ensure the long life of the whole. Sometimes technical wear and tear force a shorter lifespan, while the presence of other sub-systems is limited by user preferences. If a building meets those requirements of possible partial renewal, the owner will receive substantial tax reductions.[*10]

This is the first time that formal legal recognition is given to the dimension of time in housing policy. The practical result, of course, is that rendering a long life to what can endure and keeping adaptable what must respond to inhabitation, makes the living cell possible again.

The Japanese law applies to all residential construction. The single house as well will benefit if sub-systems can be renewed and replaced with minimum disruption of more durable parts of the building. That is not now the case. The numerous utilities that have been added to buildings in modern times have led to a notorious entanglement of subsystems including piping and wiring. Although the free standing house is the prototype living cell, this entanglement renders its metabolism sub optimal. Here as well, the new laws requirements will have a beneficial effect.

研究開発が進むと思われる。
　一つは、すでに商業施設で見られるようにスケルトンが個々のテナント店舗から切り離されたので、長寿命化し易くなったことだ。これは都市環境において道路が建物より長寿命なのと似ている。したがって望ましいパブリックスペースや建物の外観に対する長期的投資も計画し易くなった。
　現在こうした考え方による非営利住宅のプロジェクトがアムステルダムで進んでいる。このプロジェクトは、一街区に7棟の"ソリッド"と呼ぶスケルトンシステムをつくり、各ソリッドは9層のフロアーで構成する。ソリッドにできるオープンペースは賃貸され、借り手は個々のインテリアを自由に扱える。ソリッドのコミュニティの意向に反さない限り、テナントは全く自由に扱えるのだ。極めてクオリティの高いファサードとアーケードとパブリックスペースへの投資は、長期的な価値を出せると考えられ、その回収が事業に計画されている。このソリッド・プロジェクトは、その革新性ゆえに懐疑的に受取られた面もあったが、すでに完成の段階にある。
　ところで日本では、建物寿命 200年を想定した長期優良住宅促進法が 2008年に国会を通過しすでに施行されている。この法律は細部にわたる技術的条件まで触れていて、建物全体の寿命の長期化を保証するためのサブシステムの取替えの条件も設定している。技術的な面として消耗や故障などは起こるには違いないが、他のサブシステムについては、ユーザーの意向を反映できるようにしている。
　つまり住宅が、そうしたリニューアルの要求に対応できるようになっていれば、建物のオーナーは税金などの優遇を受けられる仕組みなのだ。
　このように、ハウジング関連法制において建築の寿命に関わることが公式に記されたのは世界でも初めてのことだと思う。居住に関わる耐久性と居住性能を長期にわたって保証するこの仕組みは、先に述べた生活細胞への回帰を促すに違いないと考えている。
　この法律はどんなタイプの住宅にも適用されることとなっていて、戸建住宅では、元々使われる建築部品やシステムが損傷あるいは陳腐化した際の取替え可能性がメリットと受取られて、すでに30万戸の適用実績がある。
　現在のところは部品類が多量に導入されて"もつれた状況"を呈している建築設備の配管や配線類の整理が進められているのだと思う。戸建住宅も生活細胞のプロトタイプだが、こうした"もつれた状況"については、むしろ適正なレベルにあることが多い。それでもこの法律は、戸建て住宅にとって大きな利益をもたらすことになると受取られているようだ。
　戸建住宅に較べて普及が遅れている共同住宅でも、大手ゼネコンが加盟する団体等での長期優良住宅普及に向けた取り組みが始められ、徐々に実績が出て来ている。（2011年度で分譲マンション27件、約5,800戸）。

■都市化の戦略-2：スケルトン・インフィル型産業化

　居住者に近いサブシステム部品をグループ化して一貫性あるシステムとし、そのシステムを使ってハウジングを展開することは、オープンビルディングのグローバル化した現在では、各国に共通な関心事になっている。その際には、スケルトンは細胞を包含するオープンスペースを提供し、インフィルはそのオープンスペースの容量（ポテンシャル）をフルに活用できるようにするのが一般的である。インフィル・システムは市場で入手できる効率の良い部品で構成するという考え方だ。
　今までの実績を見ると、インフィル部品の建て込みは 2〜3名の多能工チームが2〜3週間あれば完了できる事例もある。またそのコストは、1住戸分のパッケージとして、その世帯の持つクルマ購入費並みで済む例もある。
　こうした住宅部品システムの産業化は、ハードウエア（部品）につい

New opportunities: a fit-out industry

Bundling user related sub-systems into a coherent "fit-out system" is one of the major objectives of the global Residential Open Building network which advocates enabling the living cell. Where the empty base building provides space for the living cell to settle in, a dedicated fit-out system utilizes the full potential of that condition. As a composite of available sub-systems it is itself able to adopt the most efficient versions that enter the market. Experience has already shown that a dwelling unit can be fitted out by a team of two or three all-round installers in about three weeks time. The costs of such a fit-out is of the order of magnitude of the cost of the households cars.

Developing a fit-out industry is not a matter of new manufacturing - the necessary hardware sub-systems being already available - but it does require sophisticated logistics and software supported pre-assembly and packaging. Well organized, such an industry will enable urban fabric to adapt smoothly and universally to more effective and less wasteful new technology.

The first initiatives to a dedicated fit-out system on a commercial basis have been taken in Japan as well. The NEXTInfill system co. aims at serving new construction as well as renovation of existing stock. Its service is used also by a new breed of developers who buy old apartments that they entirely clean out and restore to then install new units for sale. [*11]

A sustainable built environment needs a fit-out industry that has access to all parts of extant fabrics. Conversely, for a fit-out industry to reach full potential, existing stock must be made receptive to fit-out. Fit out industry can also offer such conversion. Its effective emergence needs support of a dedicated policy including legal and fiscal initiatives.

Professional attitude and political will

For environment to be sustainable, two major conditions must be met:

Firstly, recognition of the time dimension must render long term what can endure and render adaptable what serves actual inhabitation.

Secondly, recognition of user responsibility is needed to give feedback for development of more efficient means of inhabitation and to receive such means by rapid distribution across an entire fabric.

Only cultivation of the living cell can provide these conditions.
This in turn demands a reorientation of professional habits and political encouragement:

A fit-out industry needs new ways of management and logistics. It also needs legal recognition and possible certification to assure dependable performance.

A scrutiny of legal and management aspects as well as financial policy is also needed to make both extant stock and new construction receptive to customized adaptation.

In other words, to begin cultivation, new hardware is not a pre-condition although it will, of course, be a result. What is needed is something more difficult than technical invention: a new professional attitude, explicit methodology and the application of political will.

てほぼ完成し、次に必要なのはロジスティックス（販売、生産、流通、施工の一貫システム）とソフトウエアのインテリジェント化だとされている。これらを整備して、プレアセンブルとパッケージ化の向上をはかる段階に来ている。

こうしたオープンビルディング型産業化がさらに進めば、冒頭に述べた都市環境の再整備や再編は円滑かつ効率良く実施できるはずだし、廃棄物も少なくて済むと考えられる。

こうした考え方によるインフィル部品システムの開発や実務展開はやはり日本で盛んである。中でも既にビジネスに展開されるエコキューブ・システムは先駆的なものだ。これは新築住宅市場に向けたハード、ソフトウエアであるだけでなく既存住宅のリニューアルやリモデリングにも適用できるシステムとなっている。これがきっかけになってデベロッパーは中古アパートを購入し、新部品を組込んだ再生住宅の販売を実施するようになって来た。

サステナブルな都市環境づくりには、都市を構成するファブリック全体に適用できるスケルトン・インフィル型産業が必要である。建築産業側でも、様々なファブリックに対応できるような産業再編に努力しなければならない。そうすることで建物の再生やコンバージョン（用途変更）も実施できるようになるはずだ。

こうした産業化では、法制度や公的助成制度を含めた目標の明確な政策的支援が急がれるのである。

■都市化の担い手の進化、政治の役割り

サステナブル社会における都市や住環境の形成にあたっては、次の2つが特に重要である：

第一は、現実の住生活を支えるインフィルとスケルトンの長寿命化とアダプタビリティー（フレキシビリティ）の重要さに関する認識である。

第二は、居住者（ユーザー）の責任についての自覚である。それが明確になることでより効果的で効率良くハウジングプロジェクトを計画できる。ユーザーの責任と役割が明確化することがこれからのハウジング・デザインと開発には不可欠なのだ。

以上の条件が整えば、様々に織りなす都市環境をサステナブルなものに転換できる。すなわち生活細胞の"培養"が出来るようになるからだ。このことは同時に、専門家の役割り・機能の再編成を促すので、政治的な支援も必要になるであろう。建築部品産業には今まで以上のマネージメントシステムとロジスティックスの改善が不可欠になり、そのための"認証"と"性能保証"のシステムが必要になることも認識されなければならない。

ハウジング活動全般に関連する法制とマネージメントシステムについて、詳細なアセスメントと事業性評価の検討を実施しなければならない。そうすることで新規建設とストック活用ハウジングにおけるカスタマイゼーションが実現できるからだ。

すなわち、都市の生活・産業・文化を持続可能とするためには"都市環境を耕す"という捉え方が必要であり、そのための新たなハードウエアを整備しなければならない。しかし同時に求められ不可欠なのは、そうした技術イノベーションに伴うハウジング関係者・専門家の意識改革であり、イノベーション・プログラムの明確化であり、政策当局の意思表明も不可欠であろう。

Sustainable Urbanism and Beyond -Rethinking Cities for the future, Edited by Tigran Haas, Rizzoli NY, 2012, page 204-207

*1 Habraken: Supports: an Alternative to Mass Housing, Urban International Press, U.K. Edited by Jonathan Teicher, Reprint of the 1972 English edition.

*2 Jean Gottmann, Megalopolis, The Urbanized Northeastern Seaboard of the United States, The Twentieth Century Fund, New York, 1961.

*3 Stephen Kendall and Jonathan Teicher, Residential Open Building, E & FN Spon, 2000.

*4 For projects of the last decade: see the booklet by prof. Jia Beishi at the Open Building website:
http://www.open-building.org/archives/booklet2_small.pdf

*5 See: Project Russia 20, The Free plan, Russia's shell-and-core apartment buildings, Bart Goldhoorn, editor, A-Fond publishers, 2001.

*6 The INO project, Bern, Switzerland, Client: Office of Properties and Buildings of the Canton Bern, Giorgio Macchi, director.

*7 For English language documentation of the Next21 project: GAJapan, 06, Jan/Feb 1994; DOMUS, 891, October 1999.

*8 MOOR Ltd., a subsidiary of the Tocoman group Ltd. Helsinki: www.moor.fi

*9 The IJburg Solids Project, by housing corporation Stadgenoot, Frank Bijdendijk, Director, Baumschlager.
Eberle, architects. http://www.Solids.nl

*10 The "Act for Promotion of Long-Life Quality Housing". Source: Prof. Kazunobu Minami, The New Japanese Housing Law to Promote the Longer Life of Housing. Paper, presented at the Open Building conference in Bilbao, Spain, 2010.

*11 From a unpublished report by architect Shinichi Chikazumi, Tokyo.

スケルトン・インフィル（Base Building / Fit Out）集合住宅
オランダのSolid社が開発、Amsterdamの港湾エリアに建つ

6-4　IBAエムシャーパークの実験

K.ガンザー　IBAエムシャーパーク開発公社社長
K. Ganser

はじめに

　持続可能な地域の開発では「地球規模で考え、地域で行動する」というお題目だけでは充分ではない。というのも地元の主体組織が束ねられてない場合もあるし、上位計画での設定が不充分な場合もあるからだ。

　オープンスペースと緑地という資源を扱いつつ慎重に構造転換していくプロジェクトが上手くいったのかどうかは、地域的フレームで見てみなければわからない。といってもその評価を地元の主体組織ができるわけではない。数あるプロジェクトを同時に、また共通な目標設定で実施しなければわからないのだ。

　広域な地域レベルでのフレーム条件は極めて重要である。後進地域のそれぞれの地元は、資源を考慮しないまま施設や施設利用に対して過大要求を出しがちであるし、これが開発面積、インフラ、施設規模それぞれについて過剰開発を推進する原動力になっている。

　IBAエムシャーパークでは1988年に、こうした意味の構造変革に関して、エコロジー的な基盤を形成しながら文化的な表情も備えるという目標をもってスタートした。エコロジー的基盤とは土地利用、施設利用、さらに水に関して循環型経済システムに転換することにほかならない。

　デュイスブルグからドルトムントにいたるエムシャー地域17都市に展開する、総数100を超える総額50億マルクをかけたプロジェクトを概観すれば、誰もがこの事業が成功だったと思うだろう。

　しかしIBAプロジェクトとは、ルール地域の一部での計画と建設の活動に過ぎず、この部分を外れると持続可能な地域発展につながらない開発がさまざまに行われたのである。

■ 都市化中核部開発の二重性

　成熟しきった先進工業国の大規模人口集中地域の周辺には中核部の負担になる地区が形成され、大抵はエコロジーとエコノミーの両面で問題が生じている。これらは個別の経済活動としてはプラスであっても、地域全体で見るとむしろマイナスになっている。従って、地域開発政策としては重点を中核部の整備に置くことが理に適っているし、また両方を同時に進めることもできない。住民数も職場も減少したルールのような地域で、小さくなった「ケーキ」を未だに大きいつもりで扱うのは、まさにおろかとしか言いようがない。IBAエムシャーパークでも中核部の整備に真剣に取組んだが、そこには二重の意味があった。一つは都市化地域の中心に建設投資を集約するという考え方であり、同時に、人口密集地域のど真ん中に大規模な公園をつくることであった。そして、これは実現できた。というのも脱産業化によって多数の土地が空き、建物を建てる目的が無くなったからだ。他の人口集中地域でこうした目標をかかげた事例はない。IBAは、こうした方法が可能であることを示したのである。おそらくこれがIBAプロジェクトの果たした地域開発政策上の最大の貢献と言ってよいだろう。

■ 成長なき時代の変革

　失業率が極めて高い地域（ルールでは数年来15％、都市部で25％に達している）では、ことさらに成長のあかしを求める。それはどんな成長であってもかまわないというほどだ。エコロジー、社会、文化の各面での影響など意にも解さないのである。経済面の関わりすら考慮しないこともある。量的拡大という意味での成長に向けた多大な努力は不成功

Für die nachhaltige Entwicklung reicht globales Denken und lokales Handeln nicht aus,denn es besteht die Gefahr, daß lokale Initiativen unverbunden bleiben und an übergeordneten Rahmenbedingungen scheitern.

Der sorgfältige Umgang mit der Ressource Raum und Wasser läßt sich mit Aussicht auf Erfolg nur im regionalen Rahmen angehen. Dies entwertet lokale Initiativen nicht, denn um eine spürbare Wirkung zu erzeugen, müssen viele Projekte zur gleichen Zeit und mit der gleichen Zielrichtung angegangen werden. Lokale Initiativen sind auch deswegen notwendig, um an konkreten Beispielen Bewußtseinsveränderungen einzuleiten.

Der regionale Rahmen ist aber deshalb so wichtig, weil bei lokaler Betrachtungsweise die Bedarfe an Bau- und Nutzflächen in Unkenntnis der Ressourcen in den benachbarten Räumen gründlich überschätzt werden, und dies ist der eigentliche Motor der landau und landab zu beobachtenden "Überentwicklung" bei Siedlungsflächen, Infrastruktur und gebäudebezogenen Nutzflächen.

Die Internationale Bauausstellung Emscher Park ist 1988 mit dem Ziel angetreten,dem Strukturwandel ein ökologisches Fundament und ein kulturelles Gesicht zu geben.Das ökologische Fundament besteht aus einem kräftigen Impuls, zur Kreislaufwirtschaft beim Siedlungsflächenverbrauch, bei der Gebäudenutzung und beim Wasser überzugehen.

Zumindest bei den über hundert IBA-Projekten in den 17 Städten des Emscherraumes von Duisburg bis nach Dortmund mit zusammen 5 Mrd. DM Investitionssumme kann man bescheinigen, daß dieses Ziel in großen Teilen gelungen ist. Man muß allerdings auch erwähnen, daß die IBA nur einen Teil der Planungs- und Bauwirklichkeit im Ruhrgebiet erfaßt und abseits der IBA-Einflußnahmen die Nichtnachhaltigkeit noch immer den Entwicklungsmodus bestimmt.

■ Doppelte Innenentwicklung

Alle großen Agglomerationen in den reif gewordenen Industriegesellschaften wachsen an ihren Rändern zu Lasten der Kerne, und dieses "Scheinwachstum " ist ökologisch und ökonomisch gleichermaßen unsinnig. Es bringt lediglich einzelwirtschaftliche Vorteile,regionalwirtschaftlich gesehen eher Nachteile. Die planungspolitische Forderung nach "Innenentwicklung" ist daher vernünftig und gleichwohl kaum durchsetzbar. In einer Region wie dem Ruhrgebiet, die deutlich Einwohner und Arbeitsplätze verliert, ist es besonders unsinnig, den kleiner werdenden "Kuchen" noch besonders breit auszutreten

Deshalb ist die IBA Emscher Park angetreten, die Innenentwicklung ernst zu nehmen und dies im doppelten Sinne, nämlich im zentralen Raum des Reviers die Baunachfrage zu konzentrieren und gleichzeitig auch noch einen großen Park in der Mitte der Agglomeration zu entwickeln. Das geht, weil durch die De-Industrialisierung so viele Flächen frei werden, so daß genügend übrig bleiben, die man nicht mehr für bauliche Zwecke benötigt. Ein solches Ziel hat sich wohl noch keine andere Agglomeration gesetzt, und　die IBA Emscher Park ist dabei nachzuweisen, daß dies geht. Das ist wohl der größte regionalpolitische Beitrag, den diese Bauausstellung vorzuweisen hat.

■ Wandel ohne Wachstum

In Regionen mit besonders hoher Arbeitslosigkeit - und im Ruhrgebiet liegt diese seit Jahren bei 15 %, in einigen Stadtteilen sogar bis zu 25 % - glaubt man, in besonderem Maße auf Wachstum angewiesen zu sein. Beinahe jede Form von Wachstum scheint da willkommen - ohne Rücksicht auf die ökologischen, sozialen, kulturellen und häufig auch wirtschaftlichen Folgen. Doch auch noch so große Anstrengungen, ein quantitatives Erweiterungswachstum herbeizuführen, bleiben erfolglos,

おわり、ロケーションの面ではかえって環境のクオリティを損なう結果にもなったのだ。したがって、停滞気味の地域人口と減少する職場数という条件下での、変革によるクオリティ向上が問題になる。いわば「成長なき変革」といってよい図式なのだ。IBAエムシャーパークでは、10年にわたって、「成長なき風土での変革を」という問いかけをし続け、クオリティが投資の障害になったり、クオリティ向上がコスト負担を招いたりする中で、クオリティへの認識が高まるように努めてきたのである。

■ 循環型経済

「成長しない時代の社会変革」という原理との関わりで特に顕著な事項は、オープンスペースを操作することによる循環型経済への移行なのである。そのためIBAエムシャーパーク事業の諸プロジェクトは、そのどれもが新規利用の際に整備の必要な「リサイクル用地」と絡められている。そうすることによって、「手つかずの土地」との関わりで開発済みの土地の割合がこれ以上拡がらないようにした。エコロジー面でのメリットについて特に説明する必要はないであろうが、経済面でも、こうした土地には利用されていた当時からの膨大なインフラストラクチャーがあり、これを新たに整備する必要がないのである。

循環型経済の原理は建物でも発揮される。その場所に住宅やオフィスあるいは工業施設がすでに過剰なほどあるならば、新しい建物を増やしても無駄である。そこでIBAエムシャーパークでは、既存の産業施設などの建屋スペースを慎ましく利用しようという需要を優先させる方針に切替え、それによる3つのメリットを求めた。

・用途変更に必要なリモデリング費用は相当する新築工事に比べて結局安いという経済性。
・既存建物の継続利用はエネルギーと資源のバランスの面で新築のエコ・ビルディングより有利。
・その上、産業社会時代の歴史のひとこまを後世に残せること。

またIBAエムシャーパーク事業では、雨水を下水に流さず降雨場所で地面に浸み込ませて流出を遅らせ、それによって地域の地下水の形成に役立て、小さな単位での水循環を図るという原則も採用している。この地域では1km²あたり2,000人もの居住密度があり、自然の水系が広域的に改造されてきたので、飲料水は遠隔地から引き込み、排水は遠くまでもって行っている。工業地帯の形成過程で進めてきたこうした「非自然化」を「再自然化」するという課題こそ、大規模人口密集地域の将来にとって重要なのである。

■ 従来と異なるプランニングの条件

比較的短い年月で、こうした包括的課題に取組むには、通常とは異ったプランニングの原則を当てはめざるをえない。通常なら、詳細な土地利用計画をつくり、それに応じたプログラムを整えてプロジェクト段階に進む。こうしたプランニングの原則の場合、それ自体複雑でありながら、ほぼ成立不可能な政治的合意を取り付けようとするため、プランニングの段階でつまずきがちである。あまりにも多くの問題を初期段階で一挙に扱うからだ。

IBAエムシャーパークの場合には、プランニング段階ではあまりガイドラインを設けず、比較的簡単な戦略を設定するにとどめて、「調整余地」を最大化するよう努めた。そうすることで、プログラムとプロジェクトの段階で飛躍できるようにし、実行可能なプロジェクトの戦略的な組み立てを可能にした。この「計画ではなくプロジェクトを」あるいは「プログラムではなく戦略を」という、通常と違うプランニングの条件について詳しく述べる必要があるかもしれない。とはいえ、この20年にわたって、連邦から各州へ、さらに広域から各市町へという、各レベル毎の全体計画と個別計画を作成してきた。すべての政策領域に関わる体系的で総合的な

schädigen die Standortqualität vor allem im Bereich der Umwelt nur erneut. Deshalb geht es um Veränderungen mit Innovationen und Qualität unter den Bedingungen schrumpfender Einwohnerzahlen und abnehmender Arbeitsplätze. Dies meint die Formel " Wandel ohne Wachstum ". Die IBA Emscher Park hat sich 10 Jahre mit der Frage herumgeschlagen, wie erzeugt man "Innovationen in einem nicht innovativen Milieu" und wie kann man Qualitätsbewußtsein erzeugen, wenn Qualität vorschnell als Investitionshemmnis oder Kostenlast begriffen wird.

■ Kreislaufwirtschaft

Eine bedeutsame Ausprägung des Prinzips " Wandel ohne Wachstum " ist der Übergang zur Kreislaufwirtschaft bei der Inanspruchnahme von Flächen. Deshalb wurden die Projekte der IBA Emscher Park ausschließlich auf Recyclingflächen umgesetzt, die für die neuen Vorhaben aufbereitet werden mußten. Damit konnte erreicht werden, daß sich der Siedlungsflächenanteil im Verhältnis zu den Freiflächen zumindest nicht weiter vergrößert.Die ökologischen Vorteile müssen hier nicht weiter erklärt werden, aber auch die ökonomischen sind offenkundig, denn diese Standorte haben ihre externe Erschließung aus der Vor-Nutzungszeit mitgebracht, müssen also nicht neu erschlossen werden.

Das Kreislaufwrtschaftsprinzip gilt aber auch für die Gebäude, denn wenn es eher zu viel Wohn- und Büroflächen und eher zu viel gewerblich nutzbare Gehäuse gibt, dann macht es eigentlich keinen Sinn, daneben neu zu bauen oder Altes abzureißen. Deshalb hat die IBA Emscher Park mit Vorrang die eher bescheidene Nachfrage nach Flächen auf bestehende Gebäudehüllen - zumeist alte Industriebauten - gelenkt, und dies mit dreifachem Vorteil:

- Die Baukosten für die Umnutzung waren in aller Regel niedriger als für vergleichbaren
- Neubau, also ein wirtschaftlicher Vorteil.
- Die Material- und Energiebilanz der Weiternutzung eines Gebäudes ist eindeutig besser als der vorbildlichste ökologisch orientierte Neubau.
- Und: Ein Stück Geschichte aus der Industriezeit konnte so nebenher für die Nachwelt erhalten bleiben.

Schließlich galt für die Projekte der IBA Emscher Park das Prinzip, daß das Regenwasser nicht mehr in den Kanal abgeleitet wird, sondern an Ort und Stelle verzögert versickert und verdunstet, um die lokale Grundwasserneubildung und die kleinräumigen Wasserkreisläufe wieder aufzubauen. Gerade in Regionen mit hohem Versiegelungsgrad bei einer Bevölkerungsdichte von 2.000 EW/qkm und mehr sind die natürlichen Gewässersysteme großräumig umgebaut, Trinkwasser wird aus entfernten Regionen herangeführt und Abwasser nach entfernten Regionen abgeleitet. Dieses desnaturierende Prinzip der industriellen Siedlungsweise zu renaturieren, gehört zu den wichtigsten Aufgaben der Zukunft in den großen Agglomerationen.

■ Ein anderes Planungsverständnis

Um diese Aufgaben in vergleichsweise kurzer Zeit und in diesem Umfang angehen zu können, mußte ein anderes Planungsprinzip gewählt werden. Üblich ist die Abfolge, einen detaillierten flächendeckenden Plan zu entwickeln, danach ein Programm zu formulieren und daraufhin die Projekte anzugehen. Dieses Planungsprinzip bleibt meist schon auf der Ebene der Planung stecken, weil es an der eigenen Komplexität und an nicht beschaffbarem politischen Konsens erstickt. Es werden zu viele Probleme ein für alle mal und auf Vorrat geregelt.

Die IBA Emscher Park hat daher an die Stelle des Plans eine vergleichsweise einfache Strategie mit weniger Leitsätzen gestellt, diese aber zur Maxime des Handeins gemacht. Daraufhin wurde die Plan- und Programmebene übersprungen und es wurden sofort Projekte angegangen - dort, wo sie möglich und in die Strategie einpaßbar waren. Es würde hier zu weit führen, dieses fundamental andere Planungsverständnis nach dem Prinzip "Projekte statt Pläne" und "Strategie statt Programme" näher zu erläutern. Aber nach mehr als 20 Jahren systematischer Überplanung aller Politikbereiche durch räumliche Pläne und Fachpläne auf allen Ebenen vom Bund über die Länder und die Regionen

プランニングが行われてきた結果、この国には「計画の山」が形成され、もはや意味のあるものを入れ込む隙もない状況なのだ。むしろ大事なのは、こうした重苦しい計画の予条件に新しいプロジェクトを合わせるばかりではなく、支援できるようにし、育むような「抜け道」を探すことである。そうすることによって革新的なプロジェクトの性格が、既存の基準によって骨抜きにならないようにするのである。これは「計画はプロジェクトから学ぶ」ということだ。つまり、各計画とその背後にある官僚主義に関わって「継続的例外措置」を講ずることなのである。

■「外からもたらせられるもの」と「競争」

政治にしても行政でも、あるいは同業組合などは、いつもかわらずじっと机に向い、思考方法や政治・経済を変更しなくてもすむように、慎重に進んでいくのが普通だ。「非公式なシンジケート」とでも言うべきこうした体制を壊して新しい方式や考えを取り入れようとすれば、「外からのもの」と「競争」への熱意をあえて表明しなければならない。

IBAエムシャーパークでは個々のプロジェクトの開発に際して「コンペティション」を原則としたが、じっと座っているところに突然「外からのもの」が襲いかかるという事態になる。今までは「アウトサイダー」だとされてきたものを急に本気で考慮しなければならなくなるのだ。伝統的であるがゆえに尊重されてきた施主と建築家の関係も断たれることになった。このIBAエムシャーパーク・プロジェクトではあえてアウトサイダー達の参加を可能にするような呼びかけを行ったのである。政治の世界や行政の分野あるいは許認可機関や大企業、さらには建築家や各種技術者に関しても、今までにないアイデアに驚かされたり、いわゆる基準が問われたりして、より良い解決策が取って代わるようになったのは素晴らしいことだった。これは私としても誇れる点だ。

総数60を越えるコンペティションに参加してくれた多数のプランナーと建築家と、その建築的価値実現のための条件を審査する立場にあった審査委員の皆さんにもここで感謝しなければならない。

■財政面の条件

プランニングの仕組みが整っても法制と財務面の条件が揃わなければ、「持続可能でない発展」を超える「循環のプロセス」も競争力を持ち得ない。これに関わる仕組みとして、ここでは次の二つを取り上げたい。

・1979年のルール地域会議の際に採択された発展計画で、すでに発展の道筋を示す仕組みが登場している。50億マルクの州の資金を当てた「土地基金」がそれである。この基金を使って民間市場では流通しない不動産であるため、適切な新規用途の見つからない工場跡地や余剰化している土地を取得できるようになった。この仕組みによって、IBA事業の始まる前から多量の有用な「リサイクル用地」が蓄積されてきており、投機目的ではない新規目的での利用が可能になっていた。この土地の50％以上を「ランドスケープパーク形成」に利用できたのである。こうした土地の買収費には譲渡益加算がない。投機圧力を免れた「土地基金」による土地利用の計画には想像以上のメリットがあった。多くの場合建築コンペや緑地計画コンペで決まるプロジェクトでは、経済性よりは環境のクオリティを優先させて計画することができ、それを実施できている。

・さらにもう一つの「諸制度の組合せ」が「水システム」を循環型経済に転換するためには不可欠だった。その一つは改正された州の雨水浸透義務と、下水と雨水排水の料金大系における料金の分割であり、それによって、地域毎の雨水の浸透と保水状況に合わせて雨水排水料金を自由に決められるようになった。これは州が財政支援する雨水浸透事業でもエムシャー下水道組合による雨水浸透事業でも適用されている。

hin zu den Gemeinden ist das "Plangebirge" in der Republik so dicht, daß eine weitere Verdichtung keinen Sinn mehr macht. Vielmehr muß nun eine "pfiffige Methode" gefunden werden, um mit List und Tücke in dieses schwerfällige System planerischer Vorgaben neue Projekte nicht nur einzupassen, sondern auch so zu beschützen und zu begleiten, daß ihre innovativen Qualitäten nicht an vermeintlichen oder tatsächlichen Vorschriften scheitern. Das bedeutet nicht mehr oder weniger, daß die "Pläne von den Projekten lernen müssen". Das fällt den Plänen und den dahinterstehenden Bürokratien mehr als schwer, sozusagen von einem konspirativen System der permanenten Ausnahmeregelungen umgarnt zu werden.

■ Fremdheit und Wettbewerbe

Eigentlich sitzen immer dieselben am Tisch in Politik, Verwaltung und Verbänden und sie wachen sorgsam darüber, daß ihre Denkstrukturen und ihre politischen und wirtschaftlichen Verbindungen nicht in Frage gestellt werden. Will man dieses "informelle Syndikat" aufbrechen, um neue Wege zu gehen und neue Ideen zu verwirklichen, dann gilt es, ein leidenschaftliches Bekenntnis zur Fremdheit und zum Wettbewerb abzugeben.

Bei der IBA Emscher Park wurden daher für die Projektentwicklung grundsätzlich Wettbewerbe gefordert, und daher saßen dann plötzlich "Fremde am Tisch". Nun mußten Außenseiter plötzlich ernst genommen werden und so manche altehrwürdige Pfründe zwischen Bauherrn und Architekt wurde gekappt. Der IBA Emscher Park wurde der Ruf zugestanden, daß Außenseitertum hoffähig zu machen. Schön zu beobachten, daß es viele Außenseiter in der Politik, in der Bürokratie, in den Genehmigungsbehörden, bei Großunternehmen, aber auch bei Architekten und Fachingenieuren gibt, die sich für eine ungewöhnliche Lösung begeistern konnten und so manchen Standard in Frage stellten und durch eine bessere fachliche Lösung ersetzten. Allen diesen mein aufrichtiges Kompliment.

Dank ist an dieser Stelle aber auch den vielen Planern und Architekten zu sagen, die sich an den über 60 Wettbewerben beteiligt haben, auch den Fachpreisrichtern, die mit ihrer Urteilsfähigkeit die Grundlage für das Ringen um Architekturqualität geschaffen haben.

■ Finanzielle Voraussetzungen

Jenseits des planerischen Instrumentariums müssen rechtliche und finanzielle Regelungen hinzutreten, ohne die der Kreislaufprozeß gegenüber der nicht nachhaltigen Entwicklung nicht konkurrenzfähig ist. Zwei dieser Instrumente sollen hervorgehoben werden:

• Bereits bei der Ruhrgebietskonferenz des Jahres 1979 wurde in das damalige Entwicklungsprogramm ein wegweisendes Instrument eingefügt: der mit 500 Mio. DM Landesmitteln dotierte Grundstücksfonds Ruhr. Mit diesem Fonds konnten nicht mehr oder nur marginal genutzte Grundstücke aufgekauft werden, die unter privatwirtschaftlichen Kalkülen nicht mobilisiert werden konnten bzw. zu nicht erwünschten Nutzungen geführt hätten. Auf diese Weise ist lange vor Beginn der IBA Emscher Park ein großer und wichtiger Bodenvorrat an "Recyclingflächen" entstanden, der nun spekulationsfrei für die neuen Zwecke disponiert werden konnte. Über 50% dieser Flächen dienten dem Aufbau von Landschaft. In diesen Fällen stand dem Ankauf so gut wie kein Veräußerungserlös gegenüber. Dieser Grundstücksfonds frei von Spekulationszwängen hatte den unschätzbaren Vorteil, daß die Planungen -meist über städtebauliche oder landschaftliche Wettbewerbe definiert - ohne wirtschaftliche Zwänge nach dem Qualitätskriterium entschieden

■ Werden konnten.

Ein anderes wichtiges /nstrumentenbünde/ war notwendig, um zur Kreis/aufwirtschaft beim Wasser übergehen zu können. Dazu gehört die Versickerungspflicht im novellierten Landeswassergesetz ebenso wie die Aufspaltung der Abwassergebühren in eine Abwasser- und in eine Regenwassergebühr, so daß bei Versickerung und lokaler Zurückhaltung eine Befreiung von der Regenwassergebühr möglich wurde. Dazu gehört schließlich ein vom Land gefördertes und vom Entwässerungsverband, der

おわりに

　ルール地方は150年にわたって環境に負担をかけながら発展してきた。第2次大戦後だけでも以前と同面積の開発が行われた。人口の伸びが1950年代には頂点に達したのだが、開発面積だけが倍増した。脱産業化の結果、ルール地方では住民が減少し、職場の数が減っていった。このプロセスはいまだに終わってはいない。NRW州の統計局は2015年には人口が450万人に減ると予測している。いつになれば安定期に入るのかが今のところ読めない。ある地域がもはや成長しないとすれば、持続可能な地域の発展に向けその土地利用の矛盾をより簡単に解決できるようにするのは当然のことである。他方、急激に成長する地域において持続可能な地域の発展がそもそも可能なのだろうか。可能だとすれば、プランニングと制度面での努力目標を安定状況への移行に置くことになろう。これを経済性と政治の面で適切な対価で実現できるのかどうかを、我々は問うことになる。一方でルール地域を例に取るとすれば、「永遠の成長」が無いのだから、成長の対価が「立地条件の良さ」を長年にわたって損なうわけで、遅かれ早かれ「意に反した」安定状況が生まれると考えられる。このように考えると「持続可能な地域の発展」にとって理論的回答などまずは存在しないことになる。合理的な検討を順次進める中から導き出すか、あるいは、徹底的な環境破壊段階の後に多大な費用をかけて「修繕」をおこなうのか、という選択だけであり、ルール地域の場合は後者なのである。

　とはいえ、ここで興味深くも面白いのは、著名な経済学者がすでに1884年に独自の形式で「持続可能な発展」に関して警鐘を鳴らしていることである。

　ものごとを自然の自在さにまかせられない世界を想像するのは決して心地良いことではない…

　とはいえ、この世界で経済と人口の伸びが無制限に進み、地上から貧困の大半が姿を消すことになる過程では、失われる自然に感謝はするが、以前に比べて良いとも幸福とも言えず、ただ単により多くの人々が生活出来るだけならば、私は、今後の世界について、必要不可欠なものに限った安定的状況で満足すべきだと考えるし、そうなることを期待する。
（J.S.ミル、政治経済学原理、1884年）

本稿は
IRPUD 99（1998）掲載の
Nachhaltige Regionalentwicklung durch die IBA Emscher Park
K. Ganserである。和文は澤田が作成した。
スケルトン・インフィル（Base Building / Fit Out）集合住宅
オランダのSolid社が開発、Amsterdamの港湾エリアに建つ

Emschergenossenschaft, getragenes Planungs- und Förderungsprogramm für die Versickerung von Regenwasser im Baubestand.

■ Nachbemerkung

　Das Ruhrgebiet ist 150 Jahre lang zu Lasten seiner Umweltqualität stark gewachsen. Allein in der Nachkriegszeit wurden soviel Siedlungsflächen in Anspruch genommen wie in allen Epochen zuvor. Die Siedlungsfläche hat sich verdoppelt, obwohl schon in den fünfziger Jahren die Spitze der Bevölkerungsentwicklung erreicht war. Seitdem verliert das Ruhrgebiet Einwohner und Arbeitsplätze als Folge der demografischen Struktur und der De-Industrialisierung. Dieser Prozeß ist noch lange nicht abgeschlossen. Das Landesamt für Statistik und Datenverarbeitung prognostiziert für das Jahr 2015 einen Einwohnerstand von 4,5 Mio. Ab wann ein stabiler Zustand erreicht sein wird, das läßt sich heute nicht absehen.

　Ohne Zweifel ist es aber förderlich für die nachhaltige regionale Entwicklung, wenn eine Region nicht mehr wächst, die Nutzungskonflikte werden dadurch einfacher lösbar. Deswegen soll abschließend die Frage aufgeworfen werden, ob nachhaltige Regionalentwicklung in stark wachsenden Regionen überhaupt möglich ist und wenn ja, dann müßte das Ziel aller planerischen und instrumentalen Bemühungen die Überführung in einen stabilen Zustand sein. Ob dies wirtschaftlich und politisch überhaupt konsensfähig gemacht werden kann, muß man wohl eher bezweifeln. Anderseits kann man am Beispiel Ruhrgebiet aber absehen, daß es ein "ewiges Wachstum" nicht gibt, weil die Kosten des Wachstums im Laufe der Zeit die Standortattraktivität so weit schädigen, daß früher oder später der stabile Zustand "wider Willen" eintritt. So betrachtet gibt es zu einer nachhaltigen Regionalentwicklung zumindest keine theoretische Alternative: Entweder sie wird aus rationalen Überlegungen heraus bewußt herbeigeführt, oder es werden nach einer Phase weitreichender Schädigungen aufwendige Reparaturarbeiten notwendig, wie am Beispiel des Ruhrgebietes zu sehen ist. Im übrigen ist es interessant und amüsant zugleich, daß bereits 1848 ein bedeutender Nationalökonom den Appell für eine nachhaltige Entwicklung auf seine Weise formuliert hat:

" Es ist nicht sehr befriedigend, wenn man sich eine Welt vorstellt, in der nichts mehr der Spontaneität der Natur überlassen ist, ...Wenn die Erde große Teile ihrer Armut verlieren muß, denen sie solche Dinge verdankt, die bei unbegrenztem Wirtschafts- und evölkerungs- wachstum von ihr verschwinden würden, und dies nur zu dem Zweck, eine größere, nicht aber auch bessere und glücklichere Bevölkerung auf ihr zu erhalten, dann kann ich um der Nachwelt willen nur hoffen, daß sie mit einem stationären Zustand zufrieden sein wird, lange ehe er ihr von den Notwendigkeiten aufgezwungen wird."
(John Stuart Mill, Prinzipien der politischen Ökonomie, 1848)

IBAエムシャーパーク方式の特徴

K. クンツマン
K. Kunzmann

英国・国土政策リンコルン研究所の報告書「都市のリサイクリング」（2004年）に掲載された「ドイツ・ルール工業地帯でのIBAエムシャーパーク体制の経験」でK. クンツマンが紹介している「旧工業地帯再生プロジェクトIBAエムシャーパークのアプローチは、従来のドイツ、アメリカのアプローチと何が違うか」についてまとめた資料である。

K. ガンザーの講演「IBAエムシャーパークの実験」の内容を実務面で補足していて参考になる。また、従来のドイツのアプローチ、米国で一般的なアプローチとの比較がなされているので、日本にとって理解されやすいと思う。（澤田）

Creative Brownfield Redevelopment:
The Experience of the IBA Emscher Park Initiative in the Ruhr in Germany
Recycling the City: The Use and Reuse of Urban Land.
Lincoln Institute of Land Policy, Cambridge, 2004, 201-217.

Table What distinguishes the IBA approach from traditional German and US approaches to brownfield redevelopment?

Features	IBA approach	Traditional German approach	Traditional US approach
Spatial scope	Regional, local and site	Local and site, some site focussed only	Site only
Approach	Holistic occasionally in disagreement with local government and local stakeholders	Comprehensive	Project centered
Leadership	Public sector led IBA Agency co-operating with local governments and regional and local public development agencies to initiate the project, then handed over to local development agency	Public sector led City government with, as a rule, a local or regional public developer	Private sector led Private developer in accordance with local government, (not always, though)
Citizen involvement	Considerable, form varies, however, with project character high in housing, low in other projects	Formalized, following established planning regulation procedures	Depending upon project
Budget	Drawn from a plethora of public sector budget lines (EU, national, state, local); some private investment	Drawn mainly from public sector budget (EU, national, state, local); some private investment	Private and banks
Project idea and content search	IBA-Agency brainstorming, or project idea submitted by local interest group to IBA, followed by international/national competition among architects and landscape planners	Local Government with local planners and architects; occasionally competitions or concepts of private investors	Feasibility study by architectural and business consultants
Planning control	Local government	Local government	Public sector
Effectiveness of planning control	High quality standards agreed upon project stakeholders, then followed by routine control	Routine control following established regulations	Low
Implementation	initial public investment of the state government, then highly incremental, responding to local interest and budget lines	Traditionally by local government with regional or local development agency and public pre-investment	Private investor, eventually with public support
Investment payback period	Long	Medium to long	Short to medium

表　工業地帯再生プロジェクトIBAエムシャーパークのアプローチは、従来のドイツと米国のアプローチとどう違うか

特色	IBAアプローチ	従来のドイツのアプローチ	従来の米国のアプローチ
計画対象	地方、自治体、用地	自治体管理対象地と当該用地。当該用地のみの場合もある	当該用地のみ
アプローチ	ホリスティック 地域の自治体、利害関係者との同意なく進められる場合がある。	総合的	プロジェクト立脚的
リーダーシップ	公共の主導 IBA公社が地域自治体および地方行政、地域公共デベロッパーと協力してプロジェクトを主導し、実施は地域の開発組織に引き渡す。	公共の主導 市議会が地域あるいは地方の公共デベロッパーと協力して進めるのが原則。	民間の主導 民間のデベロッパーが地域自治体と協調して開発を主導する。ただしいつもそうとは限らない。
市民の参画	重要。形式は様々。ハウジングでは参画度が高く、非ハウジングでは低い。	定められた手続きによる（縦覧）。都市計画法に定められた形式に従う。	プロジェクトごとに様々。
プロジェクト財源	過大ともいえる公共資金（EU、連邦政府、州政府、地方自治体）があり、それに民間資金が加わるかたち。	EU、連邦政府、州政府、地方自治体からの資金が主。それに民間の投資が加わる。	民間投資、銀行融資
プロジェクトの構想・コンセプト、内容の検討	IBA公社メンバーのブレーンストーミングあるいは地域の関心を持つグループが構想を提出し、国際または国内コンペティション（建築家、ランドスケーププランナー）により計画案を選ぶ。	地域の自治体が地域のプランナーや建築家の強力を得て計画案を作成。場合によりコンペティションを行ったり、民間投資の可能性を探る。	建築家とビジネスコンサルタントによるフィージビリティ・スタディで決定。
プランニングのコントロール	地域の自治体	地域の自治体	公共的組織
プランニング・コントロールの効果	高いクオリティのプロジェクトについて、利害関係者（ステークホルダー）が同意した上で、実施される。	従来からの基準に従って実施される。	効果は低い
プロジェクトの実施	州政府の公共投資を軸とし、相当量の地域の投資意欲と助成金で補充して判断する。	従来は、地域自治体が地方政府ないし地域の開発組織（デベロッパー）と公共的再投資組織と協議決定する。	民間投資家の判断。公共融資を含むこともある。
投資の回収見通し	長期	中期ないし長期	短期ないし中期

クンツマンは、以上の比較対象的なまとめの後に、IBAアプローチは紋切型でなく、基本的ルールを守りながらプロジェクトに関与する人や組織が試行錯誤していることを伝えている。

When searching for new regional development strategies for other regions in North Rhine-Westphalia, the IBA Emscher Park experience was used as a pattern for strengthening regional profiles and programming endogenous cultural development. Most sub-regions in the state pursue and apply regional modernization policies with respective lessons learn from the IBA , though with much less public financial support. In Eastern Germany, where after reunification old industrial regions are struggling for survival in a competitive world, the IBA Emscher Park Initiative has found much interest. One particular region, the former open cast lignite mining region Oberlausitz is using the approach as a sheet anchor and a straw of hope. Following the IBA model, the State of Saxony has initiated the IBA Fürst-Pückler.Land 2000-2010 , as a workshop for new landscapes (Kuhn 2000). Italian planners, informed by a special edition of the Journal Rassegna and an exhibition of the IBA at the occasion of the Architecture Biennale Venice in 1996 which had caused the publication of numerous articles in Italian architectural and planning journals, were excited by the flexible and creative approach to regional revitalization (Rassegna 1990, Wachten 1996).

In fact, hardly any other German planning achievement has attracted so much interest outside tzhe country in recent years, as the IBA Emscher Park Initiative. It is only paralleled by the Berlin project, the ambitious physical reconstruction of Berlin as a capital city. The professional echo in the United States has been less enthusiastic. Professional and political audiences, whether they come from Detroit, Pittsburgh or Buffalo, at a first glance, are excited when they see the impressive flagship projects of the IBA Initiative. However, upon further consideration they are skeptical, as to whether the visionary incremental approach would meet U.S regulation requirements, be accepted by US insurance companies and appeal to private developers in the US focussing on short term profits. However, they usually concede that the study of the many projects of the IBA initiative, and their implementation processes is a rich source of inspiring ideas for creative brownfield redevelopment, at least at such sites, where, in the absence if market pressure, brownfields in large urban agglomerations wait for unusual ideas and approaches.

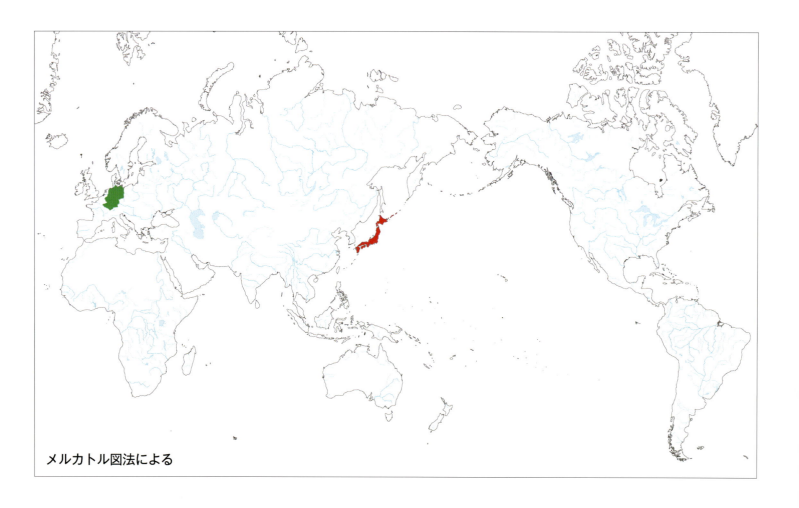

メルカトル図法による

謝辞に代えて

　本書をまとめてみて、この半世紀あまりの間にまちづくりの研究者ばかりでなく、実務に携わる人々からも多くを学んだことを思い出して、感慨深いものがある。

　ヨーロッパに関する学習は1969年のミュンヘン滞在から始まった。その際の日常生活の手ほどきは早大・吉阪研にいたシュパイデル氏から受けた。その後シュトットガルトに移ると、環境計画研究所・副所長のヴォラシェク氏（チェコの農業経済学者）との出会いがあり、ヨーロッパにおけるドイツの位置の測り方や国土・地域計画の発展経緯について教えられた。

　ヨーロッパから帰る途中、バークレーで木川田氏を訪ねたが、彼からはさらに広い視野で世界を俯瞰することに関心を持っていた。同世代の彼はオハイオで高校時代を過ごし、UCLAでは旧約聖書の世界を研究中だったから、この文化史的対話は当時の自分にとって新鮮でもあり、また刺激的だった。

　この2年の間の体験学習が、その後"産業革命後の住宅研究"を進める際の下敷きになったのだと今にして思う。

　本書の主題の一つである"建物はその造られ方を含めて捉える"という仮説は、同じ時期に、大学の恩師の内田氏とオランダのハブラーケン氏とから授かった。二人は当時それぞれ日本とオランダとで、マスハウジングに取組みながら、次の豊かさの時代に向けたハウジングの転換を考えていた。この革新には、包括的な見方が不可欠だと考え、多分野の関係者の意見を誘導して解決に導く役割を担っていた。またその検証作業を徹底して現場で行うという進め方も自分には参考になった。

　80年代に入り環境やエネルギーが問題になると、あらためて都市計画を勉強することになった。当時ようやくゼネコンに職を得て余裕もでき、伊藤滋氏や蓑原敬氏などから都市計画の概要を学ぶことが出来た。またそうして得た知識を自分のドイツでの生活体験に照らして的確に理解できるようになった。

　90年代に入り東西冷戦が終結すると、ヨーロッパ社会は一大変革期に入る。日本の関わり方も急激に変化し始めた。この時浮かび上がってきたのが旧東独・ライネフェルデの団地再生とルール工業地帯の環境再生という"まちづくりプロジェクト"である。これらのプロジェクトの課題と解決策を、日本から見た際に、どのように捉えるのが適切なのかについては、70年代からの交友関係から様々な示唆が得られた。ライネフェルデ・プロジェクトでは、ラインハルト市長自身が旧体制時の自治体間交流に詳しい人であったし、まちづくり専門家のシュトレープ氏も旧知のジーバーツ氏の門下生であった。また、ルールのプロジェクトのリーダー・ガンザー氏はシュトットガルト時代からの友人・ヴォラシェク氏の友人なので、その役割について詳しく聞くことができた。さらにこのプロジェクトをアカデミアの立場から支援したクンツマン氏とは自分が70年代にドルトムントに滞在して以来の関係である。

　「サステナブル社会のまちづくり」国際シンポジュウムは、以上のような時代を超える交流関係を基にして計画したものである。その成果を本書にまとめることが出来たのは、こうした交流の賜物だと思う。ここに名を上げて感謝の気持ちを表すと共に、こうした交流が次の世代につながることを期待する。

澤田　誠二

Acknowledgements

On completion of the writing of this book, I remember fondly what I have learned in the last half century from researchers of "Machidukuri (community development)" as well as from the people involved in its practical application.

I began learning about Europe from the time of my stay in Munich in 1969. When in Munich, I was assisted greatly in terms of daily life by Dr.Speidel, who had been in the Yoshizaka Lab at Waseda University. After I moved to Stuttgart, I met Dr.Vorazek, a Czech agricultural economist and Deputy Director of the Environmental Planning Institute. From him, I learned about the development of geographic measurements and national land and regional planning in Germany, within the European framework.

On my way back from Europe to Japan, I visited Dr.Kikawada in Berkeley in the United States. The cultural exchange I had with him and his broad outlook on the world was fresh for me at the time, and very exciting.

Thinking back, this hands-on learning during these two years then became the basis for creating "Housing Research after the Industrial Revolution."

One of the main themes of the book is the hypothesis that "buildings must be understood together with how they were built" is something that I was taught from Dr.Uchida, my university teacher, and Dr.Habraken of the Netherlands. At the time, both were studying mass housing in Japan and the Netherlands respectively, while thinking forward towards the transformation of housing in the coming era of affluence. The two believed that this innovation required a comprehensive view, and were playing the role of guiding multidisciplinary opinions from multiple specialists towards a comprehensive solution. Their work style, carried out in the field with thorough verification, served as a reference for me on how to proceed with my work.

In the 1980s, when the environment and energy became a problem, I started to study urban planning again. Here, I learned an overview of urban planning from Mr.Ito and Mr.Minohara. I was able to understand the knowledge obtained here in the proper context by comparing it against my life experiences in Germany.

In the 1990s, the Cold War ended, and European society entered a time of big changes. Japan's way of relating to Europe also began to change rapidly. During this time, "Machidukuri" projects such as housing estates refurbishment in Reinfeld in the former GDR and environmental regeneration of the Ruhr industrial region began. I was able to hear various opinions about this from my friends from the 1970s. Mr.Straeb, an expert in "Machidukuri," was also a student of my old friend Dr.Sieverts. In addition, the leader of the Ruhr project Mr.Ganser was a friend of Dr.Vorazek, a friend of mine from my Stuttgart days, so I was able to hear in detail about his role within the project. Furthermore, Dr.Kunzmann, who supported this project from the academic perspective, was a friend of mine from my time in Dortmund in the 1970s.

The "Sustainable Social Machidukuri" International Symposium has been planned based on such exchanges over the years, as described above. I would like to therefore express my feelings of gratitude to the many contributors by name, and hope that these exchanges can be continued into the next generation.

SAWADA, Seiji